THE
Mammal
SOCIETY

How to
Find & Identify Mammals

By
Gillie Sargent and Pat Morris

Contents

Page

Introduction

Records of mammals are vitally important for their conservation; we can only protect a species effectively if we know where it occurs. Fortunately, mammal records can be collected in many ways: making a sighting, finding signs, hearing a call or even by smell and their terrestrial habits also leave plenty of clues which indicate their presence. Learn how to recognise droppings, tracks, feeding remains, smells and other signs and you will be able to identify the presence of the animals that made them. This manual aims to provide the knowledge and skills to help you do this.

Your records can be used to compile atlases, which provide a baseline against which changes in distribution can be monitored, at both a local and national level. Inventories of mammals at specific sites are also needed to prevent inappropriate building developments in areas that are particularly important to mammals. Some species, such as the weasel, the mole and the pygmy shrew, are thought to be relatively common but are rarely seen. This is a dangerous situation because we may not be in a position to notice changes in their distribution. Records are also used to compile red data books allowing the relative threats to mammals to be assessed and appropriate action to be taken.

In 1995 the C.C.B.R. (Coordinating Commission for Biological Recording) reported that 65% of the biological records held in the U.K. were for birds, whereas mammals represented only 0.5%. The elusive and nocturnal habits of most wild mammals do not allow us to watch them as easily as many species of birds but this makes it very rewarding when we do catch a glimpse of one. Mammal observers who send in records can quickly make a significant contribution to the conservation of mammals by improving understanding of their distribution. We face a major challenge to put mammals on the map in the same way as it has been done for birds.

This manual aims to help surveyors improve their mammal field craft and hence their chances of making sightings and finding signs of wild land mammals. Bats and cetaceans are considered in other texts. Details are given on what to look for, where to look, how to examine the evidence and what to do with the information. It aims to help surveyors distinguish particular groups of species which often prove difficult to identify.

When looking for field signs or carrying out any form of fieldwork your own safety should be paramount. You should also ensure that you are not disturbing mammals or their habitat. Many species are protected by law and disturbance can be an offence.

Lastly, please ensure that your records go further than your notebook. Always send them to the County Mammal Recorder and help to increase our knowledge and understanding of the conservation needs of British mammals.

Part I   **Planning mammal surveys**

Why conduct mammal surveys?

Mammal recording allows us to improve our knowledge and understanding of British mammals and take part in conservation activities at a local and national level. It is also interesting to know what species live locally and fun to go and find out.

Accurate information about distributions is essential for:

❑ species and habitat site protection. Key sites might include those supporting rare species (e.g. dormice) or a large variety of species

❑ monitoring changes in distributions of species such as those likely to be important to crops (e.g. deer and rabbits) and for those affected by disease (e.g. rabbits)

❑ indicating where future survey effort is required (e.g. areas with few records)

❑ indicating where future reintroduction programmes may be successful (e.g. past records)

The best way to collate information for the above tasks is to compile local and national atlases. Mammals are difficult to study and a comprehensive county atlas usually takes several years to compile.

Do we need to record all wild mammals? Yes!

❑ **Abundant species**

Species such as the field vole, which appear to be abundant, should be closely monitored for signs of decline to allow appropriate conservation action to be taken before it is too late. Field voles provide a major food resource for many predators. A change in their numbers or distribution would have a profound effect on a number of carnivores, owls and predatory birds.

Other species may be increasing and affecting other mammals by predating on them (e.g. mink predation on water voles) or competing for their food supply (e.g. grey squirrels vs. red squirrels).

❑ **Species showing no change in status**

Surveying these allows conservationists to decide where conservation resources should be allocated.

❑ **Rare and endangered species**

Focus conservation effort where it is needed (e.g. habitat management / protection).

Case study I   **Case examples where mammal recording has contributed to conservation**

Importance of monitoring changes in range/abundance

1. A literature search of changes in the use of the word 'common' with reference to water vole abundance, led to early concern about the status of the water vole.

   Jefferies, D.J., Morris, P. A. and *Mulleneux, J. E. 1989*

2. Subsequent surveys of the distribution of water voles funded by the Vincent Wildlife Trust have alerted the conservation community to the reduction in water vole distribution and stimulated research and action to reverse the decline.

   Pat Morris

Importance of keeping inventories of threatened sites

Badger sett records are used to influence planning enquiries in well over 100 cases in Britain a year, mostly housing developments. This ensures that provision is made by developers to avoid unnecessary destruction of setts. There is an increasing trend in developments being modified to avoid disturbing the badger sett.

Pat Williams, National Federation of Badger Groups

The range of British mammals - what's out there

Britain's terrestrial species of mammal are listed in TABLE 1. They are relatively few in number compared to the diversity of mammal species on the other side of the English channel and few compared to the number of bird species on either side of the Channel.

Our 'impoverished' mammal fauna can be largely explained by climatic changes, which occurred around 12,000 years ago. As temperatures rose at the end of the last ice age Britain's endemic mammal species arrived, having migrated north from southern Europe, with 'cold adapted' species (such as the stoat) arriving first. As ice caps melted, sea levels rose separating Britain from the rest of Europe. From then on, although the climate continued warm up, the English Channel acted as a barrier preventing other terrestrial mammal species migrating north into Britain.

If you aim to detect the all of species of mammal in a given area you must diversify your methods of detection and also search in every habitat type. In some cases tracking a mammal may involve a variety of search methods looking for, say, droppings, feeding remains and, perhaps, a fleeting glimpse of the animal itself. All this evidence adds up to a stronger indication that a given species is present.

Bats and cetaceans (whales and dolphins) which require special techniques are considered elsewhere.

TABLE 1 Mammals (other than bats) occurring on mainland Britain and their usual habitats

Many species listed favour habitat diversity.

	Woodland	Riverside	Urban Garden	Mountain Moorland	Farmland	Grassland
Hedgehog	❏		❏		❏	❏
Mole	❏	❏	❏		❏	❏
Common Shrew	❏		❏	❏	❏	❏
Pygmy shrew	❏		❏	❏	❏	❏
Water shrew	❏	❏			❏	
Rabbit	❏		❏	❏	❏	❏
Brown hare	❏				❏	❏
Mountain hare				❏		
Red squirrel	❏		❏			
Grey squirrel	❏		❏		❏	
Bank vole	❏		❏		❏	❏
Field vole			❏	❏	❏	❏
Water vole		❏				
Wood mouse	❏		❏		❏	❏
Yellow-necked mouse	❏		❏			
Harvest mouse		❏	❏		❏	❏
House mouse			❏		❏	
Brown rat		❏	❏		❏	❏
Black rat		Dock side only ❏	Buildings near dock ❏ sides only			
Hazel dormouse	❏					
Fat dormouse	❏		❏			
Fox	❏		❏	❏	❏	❏
Pine marten	❏			❏		
Stoat	❏			❏	❏	❏
Weasel	❏	❏			❏	❏
Polecat	❏	❏		❏	❏	❏
Feral ferret				❏	❏	
Mink		❏			❏	
Badger	❏		❏	❏	❏	
Otter		❏				
Wildcat	❏			❏		
Red deer	❏			❏		❏
Sika deer	❏					
Fallow deer	❏				❏	
Roe deer	❏			❏	❏	❏
Muntjac	❏		❏		❏	❏
Chinese water deer	❏ & reed beds					
Number of species	**27**	**10**	**19**	**14**	**24**	**18**

Where and when to conduct mammal surveys

Mammals are elusive, but chance sightings occur with patience and luck. Knowing where best to look for the signs (e.g. under bridges for otter spraint) saves a lot of time, and notes on this are given later under the sections on signs. Certain habitats are favoured by particular species and it may be easier to locate signs in particular habitats (see TABLE 1). Although there is little point in spending considerable amounts of time looking for species that have never been recorded in an area, it is also worth keeping an open mind. It is particularly important to try to record mammals in areas that have been previously under-recorded. Checking discarded bottles, for example, is a particularly good method for collecting records in inaccessible, under-recorded locations such as mountainous landscapes where little trapping is likely to have been done in the past.

Nocturnal mammals tend to be most active at dusk and dawn so these are the best time to be out looking for them. Ideally there should still be enough light to observe signs. At dawn signs left overnight are more likely to be fresh. Use the signs mammals leave as clues to lead you to a particular point and to improve your chance of making a sighting. TABLE 2 gives a calendar for conducting surveys for sightings and signs.

Table 2 Mammal survey seasons - best times for success

Spring	Summer	Autumn	Winter
Hare, squirrel and other species show breeding activity; defending territories and raising young. Hares are easier to see before the grass and crops grow tall.	Courting adult hedgehogs can be heard 'puffing' and are easily spotted by searching open areas with a torch. Water vole latrines are produced from spring to summer. Fat dormice can be heard making calls. Badger - good time for seeing young playful animals. Feeding areas, runs and latrines become more obvious as the animals are more active. Shorter nights and the need to find food for young force many species (e.g. fox and badger) to be active in daylight hours which may result in more sightings.	Hedgehogs may be observed preparing for hibernation in piles of fallen leaves. Some small mammals; bank vole, wood mouse, hazel dormouse and squirrels accumulate nuts at feeding sites; the forager can be distinguished by characteristic teeth marks (see FEEDING REMAINS). Harvest mouse nests and squirrel dreys become more conspicuous as vegetation dies back. They begin to disintegrate, however, in early spring or after strong winds. Deer can be seen and heard rutting especially during September and October. Relatively large populations of all species have dispersing young which results in more sightings in autumn. There is often much feeding activity before the first frosts.	Yellow-necked and wood mice often inhabit buildings in winter. Scarce food may force species that remain active in winter to feed in more open situations. Snow will reveal footprints (see Table for species that can be identified easily from tracks).

What to take on a mammal search

The equipment you need depends on the survey work being carried out, but certain items are worth keeping handy at all times when you are out and about.

1. Maps (if in doubt about how to take a grid reference see APPENDIX 1).
2. Collecting containers such as plastic pill boxes, bottles for droppings (e.g. 25ml film canisters), bags for carcasses (if you want to examine these more carefully at home or keep a reference collection).
3. Rubber gloves for handling carcasses or droppings (see APPENDIX 3 on health and safety).
4. Compass for determining position on map.
5. Recording forms and code card.
6. Binoculars (always carry binoculars in case you see a mammal in the distance, or want to look more closely at a sign in an inaccessible place such as a water vole latrine on the other side of a river).

Wear clothes of dark or muted colours and nothing that rustles or jangles. Once you are in sight of a mammal try not to move, but if you must, do it slowly. As most mammals have a keen sense of smell avoid wearing perfume or aftershave and try, if possible, to position yourself downwind of the mammal.

Asking around

Much time can be saved by talking to local people whose daily lives may involve encounters with wild mammals e.g. gamekeepers, foresters, farmers and pest controllers. It may be useful to have a list handy of species appropriate to the county with a list of some of the other names sometimes used. You may find it useful to circulate a simplified questionnaire or recording form to target certain sets of people such as fishing clubs or gamekeepers or use the local newspaper or radio station.

Reports of species which are familiar to most people such as hedgehog, fox and badger can generally be taken at face value. Reports of other species such as dormouse and water vole may need following up. Stoat and weasel are often confused, as are mink and polecat. You need to enable the County Mammal Recorder to judge the competence of the observer if passing on records from other people. If you are in doubt about the validity of the record try to confirm it yourself by using the information you have been given, to direct you to the location sighting or other evidence.

When making the record yourself try to be as precise as possible. "Joe Bloggs saw a stoat on 3rd July 1995 at SN123456" is more useful than "J. Bloggs saw a mustelid in summer 1995 near the station". This makes data analysis much easier and may be more useful for protecting a particularly important site from a damaging planning application.

Case study 2   Local questionnaire requesting local information on a species

A water vole survey was carried out by Rob Strachan and the Wildlife Conservation Research Unit (WildCRU) by placing posters around Oxford city and inviting local people to a water vole survey training day. Sixty attended and 75% of the forms were returned identifying 5 new water vole colonies.

Rob Strachan

Part 2 Making sightings

This section is divided into small mammals (head and body up to 130mm) and large mammals (head and body more than 130mm), which is further divided into herbivores and carnivores. Groups of species that can be difficult to distinguish are considered on the same page. Species that most people are familiar with, such as fox and badger, are not discussed. Where drawings are not to scale a picture of a standard object has been included on each of these pages to indicate the relative size of the animals.

Small mammals - head and body less than 130mm long

Small mammals are easy to identify providing you remember a few key features associated with each species.

Common shrew, pygmy shrew and water shrew

The water shrew can be distinguished from the other two species by its larger size and its habit of diving and swimming as well as by the features mentioned below.

WATER SHREW

APPROXIMATE
ACTUAL SIZE

PYGMY SHREW

COMMON SHREW

Features	Common shrew	Pygmy shrew	Water shrew
Coat colour	Dark brown—three tone coat; dark back, paler sides and paler still underside	Paler brown—two tone coat; dark back and pale underside	Black fur on back and silvery grey underside with sharp demarkation in adults
Tail	A few stiff hairs under tail	A proportionately longer, thicker and more hairy tail. A few stiff hairs under tail	Prominent keel of stiff silvery hair under tail like a rudder
Feet	Short hairs fringing toes	Short hairs fringing toes	Prominent stiff silvery hairs fringing toes
Head	Not particularly domed	Domed	Not particularly domed
Eye patch	No eye patch	No eye patch	Often with white patch above eye
Distinctive features			Larger with a habit of swimming and diving

Wood mouse, yellow-necked mouse, house mouse, harvest mouse and hazel dormouse

Adult yellow-necked mice are 1.5 times bigger than an adult wood mice, but can only be distinguished for sure by examining the underside of the neck in the hand. Yellow-necked mice have an unbroken yellow band passing across the chest, linking the forelegs, whereas wood mice have a white chest usually with a longitudinal orange streak or spot. Mind your fingers! Yellow-necked mice tend to be more exuberant and bite more readily (and harder) than wood mice. Young house mice and wood mice are rather similar, although house mice are always grey-brown with smaller eyes and ears and a thicker hairless tail. Wood mice have a white underside and a yellow brown back, whereas house mice are all one colour though slightly darker on top. House mice have greasy fur and smell strongly. The hazel dormouse can be identified from true mice, when adult, by its furry tail and orange yellow fur.

Features	Wood mouse	Yellow-necked mouse	Harvest mouse	House mouse	Hazel dormouse
Chest marks	White chest and usually a longitudinal orange patch or streak between forelegs	Unbroken yellow band passing across the chest lining the forelegs	None	None	None
Colour of back	Reddish-brown	Reddish-brown	Golden-brown	Greyish-brown	Adults orange-brown Juveniles greyish-brown
Colour of underside	Pale grey/white	Very pale grey	Very pale grey	Greyish-brown	Pale buff to white on throat
Ear size	Large (15--17mm)	Large (16--18mm)	Small (7--9mm)	Medium (12--15mm)	Medium (about 12mm)
Tail	Fur black on top	Fur black on top	Prehensile, brown/pink all over with some fur	Brown/pink all over with some fur	Bushy
Other distinctive features			Smaller than other British mice with a blunt, vole-like muzzle	Greasy fur and strong 'mouse' smell	Short muzzle, long black whiskers and prominent black eyes

Viewing tips

To attract small mammals try placing seeds, raisins or chocolate drops for rodents, and casters (fly pupae available from fishing shops) for shrews, at an observation point, such as in front of a window or on a mammal table, on a regular basis. Put the seeds out just before dusk to prevent them being taken by birds. Put a cover over the area if small mammals are at risk of being taken by predators at the mammal table.

APPROXIMATE
ACTUAL SIZE

YELLOW-NECKED MOUSE

WOOD MOUSE

HOUSE MOUSE

DORMOUSE

HARVEST MOUSE

Field vole and bank vole

Features	Bank vole	Field vole	Orkney/Guernsey vole
Colour of back	Reddish brown	Yellowish or greyish brown	Yellowish or greyish brown
Tail length as % of head and body	50	30	30
Tail Colour	Dark on top, white below	Pale brown all over	
Ears	Prominent	Hardly visible	

BANK VOLE

FIELD VOLE

APPROXIMATE
ACTUAL SIZE

How to Find and Identify Mammals
The Mammal Society

Figure 1 **British small mammals exaggerating distinguishing features**

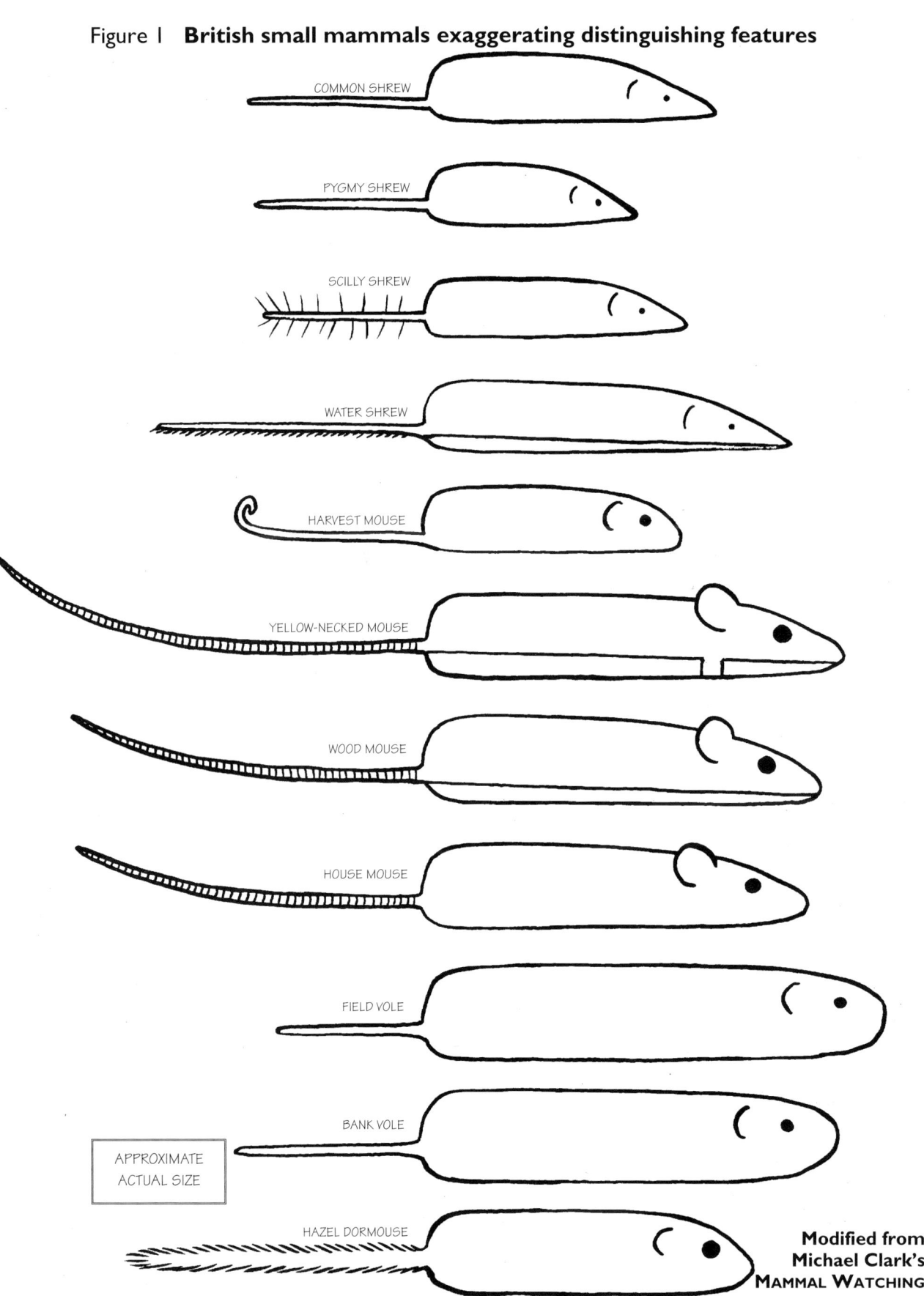

COMMON SHREW

PYGMY SHREW

SCILLY SHREW

WATER SHREW

HARVEST MOUSE

YELLOW-NECKED MOUSE

WOOD MOUSE

HOUSE MOUSE

FIELD VOLE

BANK VOLE

APPROXIMATE
ACTUAL SIZE

HAZEL DORMOUSE

**Modified from
Michael Clark's
MAMMAL WATCHING**

Large mammals (Herbivores) - head and body more than 130mm long

Fat dormouse, grey squirrel and red squirrel

An adult fat dormouse is actually hard to confuse with a hazel dormouse! It is twice the size and grey. It is more easily confused with a grey squirrel because of its colour and bushy tail. The tail of the fat dormouse is, however, dark brown whereas the grey squirrel's is khaki with a white fringe.

Viewing tips

In urban settings both red and grey squirrels often become tame and will even feed from the hand. In rural surroundings, however, both species tend to remain shy. Greys tend to spend more time on the ground whereas reds spend nearly all their time in the tree canopy and are harder to spot. Red squirrels are probably best spotted by walking slowly along forest paths looking up into the canopy. In winter reds may become tinged with grey but develop characteristic ear tufts which distinguish them from greys. In summer the grey squirrel can have very brown feet and flanks.

Features	Dormouse	Fat dormouse	Grey squirrel	Red squirrel
Tail	Bushy - orange-brown	Very bushy grey tail with all the hairs a single colour	Tail hairs are banded in brown, black and yellow, creating a multi-coloured tail	Bushy and red (all one colour) and occasionally very dark Pale in summer
Coat	Orange-brown	Grey	Grey - often with brown tinge on back, especially in summer	Reddish-brown (grey tinge in winter)

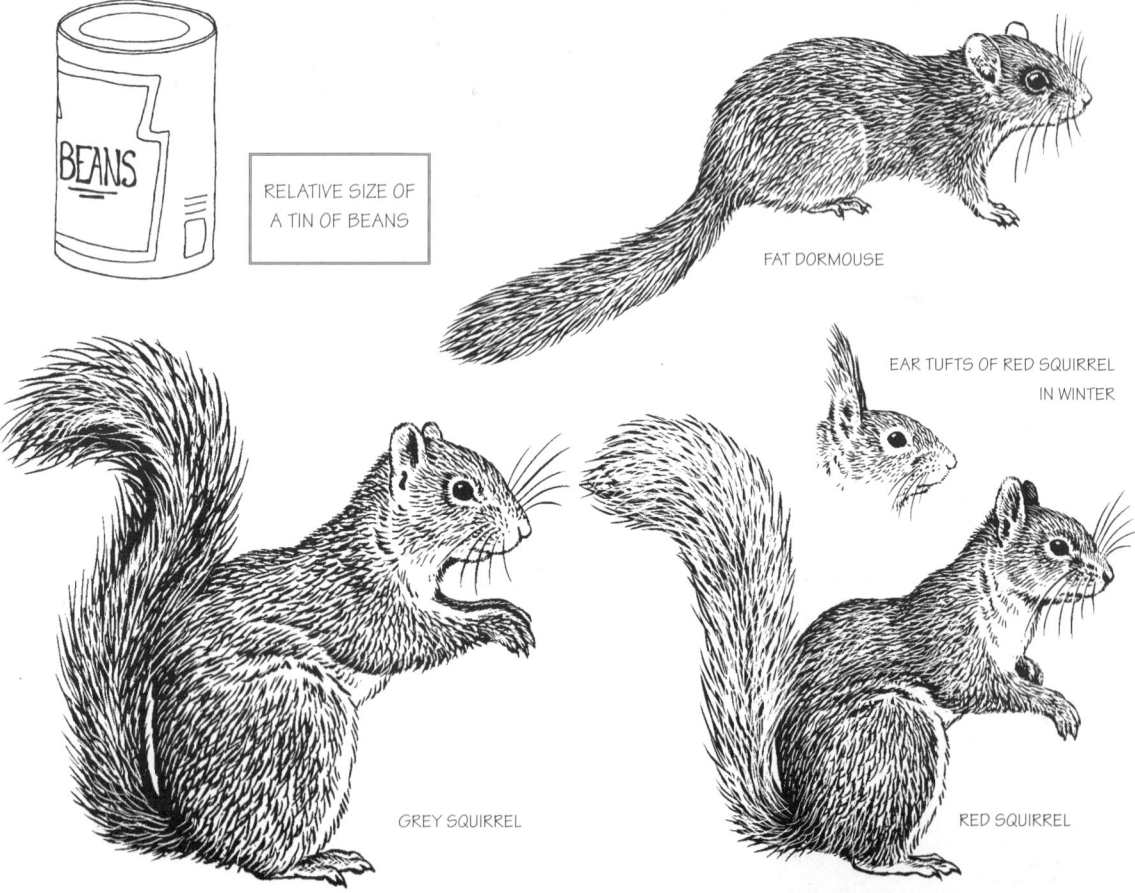

BEANS

RELATIVE SIZE OF A TIN OF BEANS

FAT DORMOUSE

EAR TUFTS OF RED SQUIRREL IN WINTER

GREY SQUIRREL

RED SQUIRREL

How to Find and Identify Mammals
The Mammal Society

Water vole, brown rat and black rat

Viewing tips Water voles can be spotted by ambling slowly along a suitable canal or river. Do not automatically assume that a swimming rat-sized animal is a water-vole; all three species are aquatic and can be observed particularly well from a vantage point such as a bridge. Once you have spotted an animal, move only when it is under water, to reduce the distance between you and the animal.

Features	Water vole	Brown rat	Black rat
Coat colour	Dark brown (may be black, especially in northern Scotland)	Grey brown	Black (longer, sleeker fur than brown rat) can sometimes be grey brown
Muzzle	Blunt and vole-like	Pointed	Very pointed
Tail	Slightly furry and much shorter than the body	Hairless and shorter than the body	Hairless and 20% longer than body (sometimes shorter)
Ears	Very small	Large	Large
Sound on entering water	A characteristic 'plop'	No sound	Doesn't enter water

BLACK RAT

WATER VOLE

BEANS

RELATIVE SIZE OF A TIN OF BEANS

BROWN RAT

Mountain hare, brown hare and rabbit

Viewing tips

To spot brown hares, which, compared to rabbits have a loping gait, scrutinise any short crops such as winter wheat. Go early in the day and use binoculars or a telescope to inspect fields. Do the same in heather moorland to improve the chances of seeing mountain hares. These species tend to be most active at dusk and dawn.

Features	Mountain hare	Brown hare	Rabbit
Ear length	Midway between brown hare and rabbit	Long - about twice length of the head	Short - about the same length as the head
Ear tip colour	Black	Black	Brown
Tail colour	All white	Black on top	Dark on top
Eyes	Brown	Golden staring eyes	Brown
Body colour	Grey (or white in winter)	Grizzled, orange-brown flanks	Greyish-brown

MOUNTAIN HARE

BROWN HARE

RELATIVE SIZE OF A TIN OF BEANS

RABBIT

How to Find and Identify Mammals
The Mammal Society

Deer - red, sika, fallow, roe, muntjac and Chinese water deer

	Red	Sika	Fallow	Roe	Muntjac	Chinese water deer
Rump	Buff	Heart shaped white with black upper border flared out when alarmed	Heart shaped white with horseshoe shaped black border	Cream-white which can be flared when alarmed	Dark	Dark
Tail	Ginger buff	White (with thin vertical black streak)	Black and comparatively long	No visible tail. In winter female has a tuft of long hair between back legs	When alarmed, tail held vertical to show white underside	Stumpy - never held erect as in muntjac
Seasonal coat change		Distinct spots in summer on brown	Distinct spots in summer. Background depends on the morph. Sometimes all dark when the spots are not visible		Chestnut brown in summer darker brown in winter	
Antlers of mature male	Large branched	Typically > 4 points per antler	Upper part of antler broad and flattened	Small branches typically not > 3 points	Single spikes pointing backwards with or without very small brow line seldon > 10 cm	None
Other distinguishing features	Coat changes from grey brown in winter to red brown in summer	Grey brown black in winter, no spots	Colour variable ranging from white to black but mostly brown with spots	Black nose and white chin	Males have tusks (upper canine teeth) protruding up to 2cm below upper lip	Males have long curved tusks (upper canine teeth) protruding up to 7cm below lower jaw
Females	Smaller than males and no antlers					

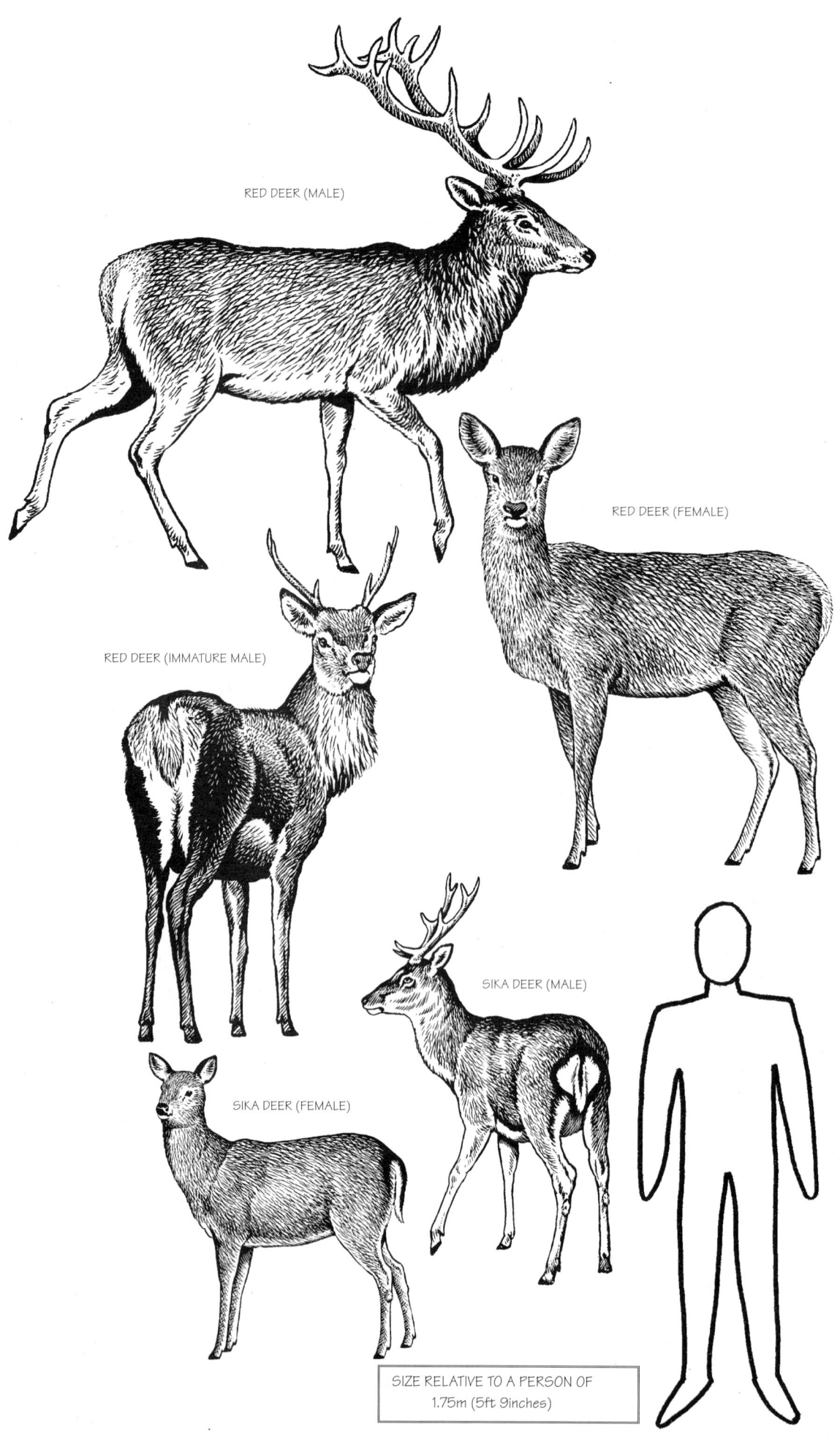

RED DEER (MALE)

RED DEER (FEMALE)

RED DEER (IMMATURE MALE)

SIKA DEER (MALE)

SIKA DEER (FEMALE)

SIZE RELATIVE TO A PERSON OF
1.75m (5ft 9inches)

How to Find and Identify Mammals
The Mammal Society

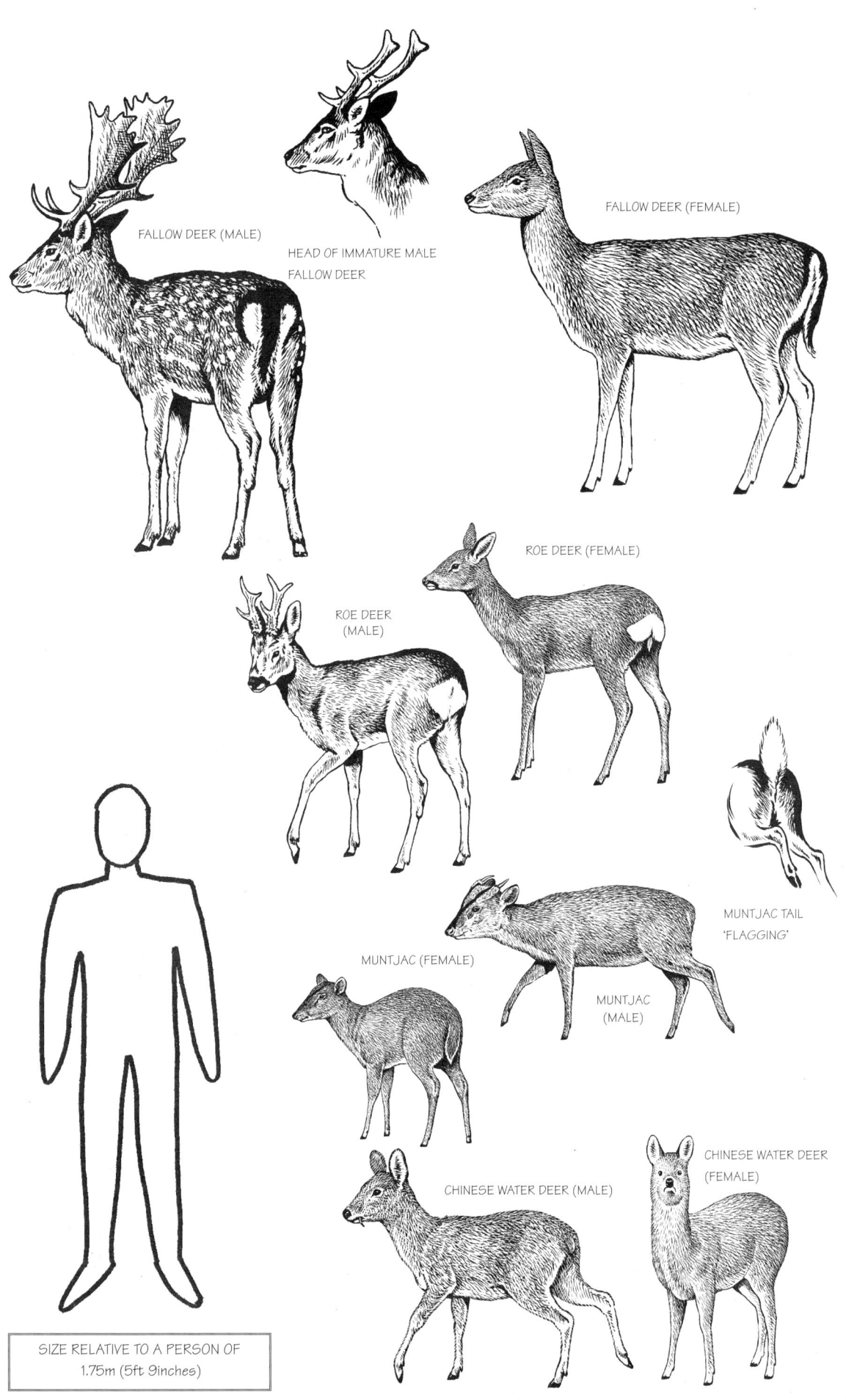

FALLOW DEER (MALE)

HEAD OF IMMATURE MALE
FALLOW DEER

FALLOW DEER (FEMALE)

ROE DEER (FEMALE)

ROE DEER
(MALE)

MUNTJAC TAIL
'FLAGGING'

MUNTJAC (FEMALE)

MUNTJAC
(MALE)

CHINESE WATER DEER
(FEMALE)

CHINESE WATER DEER (MALE)

SIZE RELATIVE TO A PERSON OF
1.75m (5ft 9inches)

Large mammals (Carnivores) - head and body more than 130mm long

Weasel and stoat

Weasels and stoats are both much smaller than other British carnivores, and are the only mustelids with a red/brown back and white underside. They are often seen during the day.

Stoats, even young ones, have a black tip to the tail (the tip being the last 20%) and this feature distinguishes them from all other British mammals, including weasels.

Features	Weasel	Stoat
Tail colour	All red-brown	Black tip
Flank colour division	Wavy	Straight (wavy in Irish stoats)
Underside	White	Creamy or pale yellow

WEASEL

STOAT

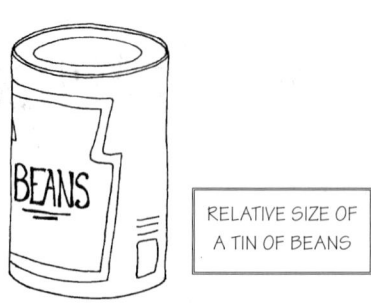

RELATIVE SIZE OF A TIN OF BEANS

How to Find and Identify Mammals
The Mammal Society

Polecat, polecat-ferret, mink and pine marten

All four species are much larger than stoats and darker brown (although feral ferrets may vary from albino to dark brown and are sometimes almost indistinguishable from wild polecats). They are also more nocturnal and harder to observe. Compared to a domestic cat all have short legs and thick tails. The ears of the pine marten are slightly more conspicuous than the polecat, polecat-ferret and mink and it has a slightly longer and bushier tail. The pine marten is the only one of these species which may be seen in trees. Apart from direct sightings most records are from road casualties.

Features	Polecat	Polecat-ferret	Mink	Pine marten
Face	White face band	White face band (usually less distinct than polecat)	All brown except a small white patch under the chin	Dark brown
Chest	Blackish	Dark brown—may have small white patches	Dark brown	Creamy white
Coat colour	Blackish with pale underfur	Blackish with pale underfur often paler than polecats & with more colour variation	Usually dark brown	Dark brown

POLECAT

POLECAT-FERRET

MINK

PINE MARTEN

RELATIVE SIZE OF A TIN OF BEANS

BEANS

Otter and mink

Features	Otter	Mink
Coat colour	Mid-brown	Chocolate brown/black
Tail	Long, tapering and sleek	Cylindrical and fluffy
Face	Broad muzzle	Pointed muzzle
Size	Larger than cat	Smaller than cat
Other features	Creates bow wave when swimming	No bow wave created

Viewing tips

An adult otter is much larger than a domestic cat whereas mink are smaller. In the water an otter can be distinguished from a mink and aquatic rodents by its large size, flattened head and the distinctive V-shaped wake.

RELATIVE SIZE OF
A TIN OF BEANS

OTTER

MINK

OTTER HEAD
(NOT TO SCALE)

OTTER SWIMMING

MINK HEAD
(NOT TO SCALE)

MINK SWIMMING

WATER VOLE SWIMMING

How to Find and Identify Mammals
The Mammal Society

Wildcat and feral cat

Wildcats produce fertile hybrids with domestic cats. There is therefore, great variation in the appearance of these cats in the wild. Specimens like the wildcat illustrated below still exist and should be carefully recorded.

Features	Wildcat	Feral cat
Tail	Thick, bushy and blunt with a black tip	Tapering tail
Colour	Tabby with flank stripes	Various

WILDCAT

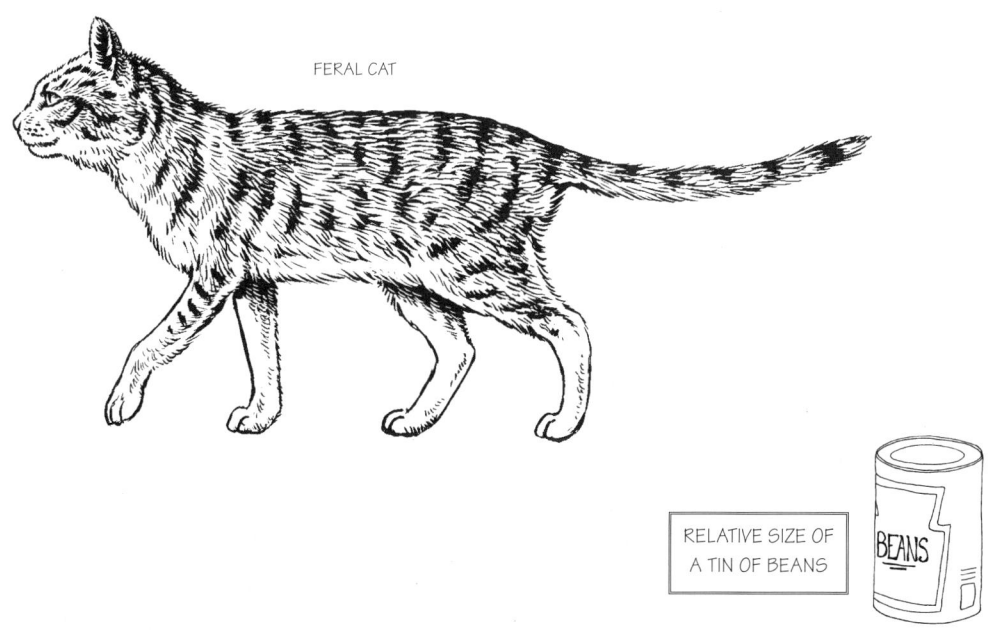

FERAL CAT

RELATIVE SIZE OF
A TIN OF BEANS

BEANS

Part 3 Identifying mammal calls

Table 3 **Identifying mammal calls**

Species	Description of calls and commonly made sounds	Season
Hedgehog	Puffing when courting	Summer
Shrew	High pitched, almost ultrasonic squeaks.	
Rabbit	Scream when attacked by a predator.	
Yellow-necked mouse	Squeak vociferously, especially when handled.	
Fat dormouse	Calls, a churring sound made from trees, can be heard up to km away on a still night.	Summer
Grey squirrel	Churring noises in trees.	
Fox	Short barks, higher pitched than dogs with screams.	Early in the year
Badger	Cubs may make a short high-pitched yelp when playing.	Spring
Otter	Whistle is made as a contact call. 'Hah!' is an anxiety call.	
Wildcat	Hiss when disturbed.	
Roe deer	Short alarm bark.	
Red deer	Stags roar at adversaries.	
Sika deer	Stags make a variety of sounds, including a distinctive whistle and a sound like a squeaky hinge.	
Fallow deer	Bucks make a loud gurgling belching noise and short alarm bark.	
Muntjac	Loud single bark when alarmed, when does are in season or have just given birth.	
Chinese Water deer	Loud dog-like bark.	

How to Find and Identify Mammals
The Mammal Society

Part 4 **Looking for signs**

Mammals, except bats, tend to spend more of their time on the ground than birds and often communicate, or mark their territory, by leaving droppings or scent marking (rather than by displaying a colourful plumage or singing like birds). Consequently, there are usually more signs of their presence in relatively accessible places. Most British mammals are small, nocturnal or both and therefore tend to escape notice unless we find them dead (eg regurgitated in owl pellets or in discarded bottles) or we use special techniques to catch them (in traps) or detect them (e.g. hair tubes). Road casualties are also an important source of records of mammals that would often remain undetected. Certain methods are more appropriate for some species than for others. Use TABLE 4 as a 'prompt list' of signs commonly left by each species.

Some signs can be taken as a direct identification of the species, for example, the spraint of an otter is almost definitive, whereas the dropping of a fallow deer can be confused with other species of deer, or perhaps sheep, except by the experienced eye.

Some signs provide clues which need to be followed up by further investigation to provide a reliable record of the species. A burrow, for example, along with a distinctive footprint beside it, may confirm what species is using the burrow. As soon as you find one sign, look for others in the immediate vicinity to back up the record. If you have been unable to record a species you believe occurs in a given area, refer to the ATLAS OF BRITISH MAMMALS (*Arnold 1994*) to see what kind of recording techniques have been successfully applied in the past.

Arrival at the site

First scan the site, a field for example, for obvious signs. Mole hills fall under this category, and possibly rabbit burrows. Try to think like a wary mammal. Hedgerows can act as corridors linking patches of denser habitat, such as woodland, with signs scattered along the way. Cross open spaces to inspect any landmark features (e.g. a gate post) for territorial markings such as droppings. (Use TABLE 4 as a prompt). Remember to record common species at least once a year for each area surveyed.

Table 4 Techniques for recording terrestrial British mammals
other than sightings of live animals

	Signs							Dead mammals			Traps			
	Burrows	Nests	Tracks/trails	Fur	Droppings/latrines	Feeding signs	Road casualties	Discarded bottles	Owl pellets	Cat kills	Longworths	Nest boxes	Corrugated iron	Hair tubes
Hedgehog		□			□		□					□		
Mole	□						□		□	□				
Common shrew								□	□	□	□		□	□
Pigmy shrew								□	□	□	□		□	□
Water shrew					□			□	□	□	□		□	□
Rabbit	□		□		□		□		□	□				
Brown hare			□		□		□							
Mountain hare			□	□	□		□							
Red squirrel		□				□	□			□		□		
Grey squirrel		□				□	□			□		□		
Bank vole		□						□	□	□	□		□	□
Field vole								□	□	□	□		□	□
Water vole	□		□		□	□		□	□					
Wood mouse								□	□	□	□	□	□	□
Yellow-necked mouse								□	□	□	□	□	□	□
Harvest mouse		□						□	□	□	□			□
House mouse								□	□	□	□			□
Brown rat			□		□		□	□	□	□			□	□
Hazel dormouse		□				□		□	□	□		□		□
Fat dormouse							□		□			□		
Fox	□		□	□	□		□							
Pine marten					□		□							
Stoat							□							
Weasel							□						□	
Polecat							□							
Mink			□		□		□							
Badger	□		□	□	□	□	□							
Otter			□		□		□							
Wildcat							□							
Red deer			□		□	□	□							
Sika deer			□		□	□	□							
Fallow deer			□		□	□	□							
Roe deer			□		□	□	□							
Muntjac					□	□	□							
Chinese water deer					□		□							

□ denotes a sign that may be distinctive enough to allow identification to species level by non-specialists. A ring indicates a group of species whose signs are easy to identify as belonging to the group but hard to identify to a particular species

Tracks and trails

Tracks or footprints are the imprint of the underside of a paw made in damp, soft substrates such as sand or mud. Both tracks and trails are often a good clue but are not generally regarded as reliable records on their own, and usually need following up to obtain a confirmed record.

Tracks

In order for clear tracks to be made that can be identified with confidence a number of conditions need to be satisfied. The substrate needs to have the right consistency and the animal must have left a clean track which has not been smudged or superimposed. It then takes an experienced eye to distinguish species which leave similar tracks, such as dog from fox print. Domestic mammal tracks account for around 90% of tracks that one encounters in the countryside, and it is important to be able to recognise these first to allow you to pick out those of wild mammals. Consider the habitat you are in when identifying tracks but keep an open mind at the same time. The pattern of tracks can also be distinctive.

The best conditions for looking for tracks occur when there is fresh snow. Admittedly, this makes your survey planning difficult, but is an opportunity worth pursuing when you can! Tracks made only a few hours earlier are easiest to identify and may tell you something about the animal's behaviour. If it was hunting, the tracks of the predator may lead to the tracks of the prey or if it was courting you may find the tracks of another individual. Bear in mind that in certain circumstances (e.g. in melting snow) the tracks will become enlarged. Tracks of cats, for example, may appear to be more like those of a puma! Tracks can also be clear in shallow mud after rain.

Tiger tracers and tracker boards

A simple method of recording tracks for later reference is to cover the tracks with an acetate (transparent plastic) sheet and draw an outline of the track using a felt tip. This method was devised by hunters to record individual Indian tigers and is known as a 'tiger tracer'.

Tracker boards consist of a tile covered in soot by holding it over a candle and can be used to detect the passage of small mammals.

Plaster casts

In firmer, deeper mud a plaster cast of tracks can be taken, especially if the species is under recorded in the area or if the species is locally or nationally rare. This allows the track to be examined in three dimensions to confirm its identity.

1. Mix water and Plaster of Paris (available from chemists) in equal parts. Add plaster to water stirring to mix well and tap the container to ensure there are no air bubbles.
2. Enclose the track with a mould such as a plastic ring, which can be made by cutting up a plastic container such as a bottle of washing-up liquid or an ice cream container, or a ring of card held in place by paper clips, and pour in the plaster.
3. Allow the plaster to dry for about 30 minutes (longer in damp weather).
4. Lift the mould, allow it to dry completely, then clean off any mud with a soft brush such as an old tooth brush, or wash it.
5. To create an imprint of the cast, cover it with a layer of petrolum jelly, place the cast in the mould by surrounding it with a strip of card to contain the wet plaster, and repeat the above procedure.

Identifying mammal tracks

In order to learn particular mammal tracks, it helps to be familiar with the structure of the mammalian foot and how it varies between species. TABLE 4 indicates species that can be easily identified by their tracks.

The structure of the foot reflects the lifestyle of the animal; fast runners such as deer run on their toes (hooves) whereas those that tend to plod, such as badgers, do so on the whole foot (the paw). Swimmers, such as otters, have webbed feet. Hooved animal tracks display a single large toe (e.g. a horse) or two slots (or cleaves) such as deer and sheep. Dew claws, which are positioned higher up the foot, may show up in deep snow. Non-hooved mammals leave an imprint showing four or five toes around a pad, the sole of the foot, and often claw marks.

Tracks

Insectivores and rodents

There is much overlap in the size and appearance of the tracks of these groups of species. They can rarely be used for positive identification to species but you may be able to identify them to their respective groups. The hind feet of both insectivores and rodents have five toes and claws, insectivores (hedgehog, mole and shrews) show five toes on their fore foot, whereas rodents show only four.

Insectivores

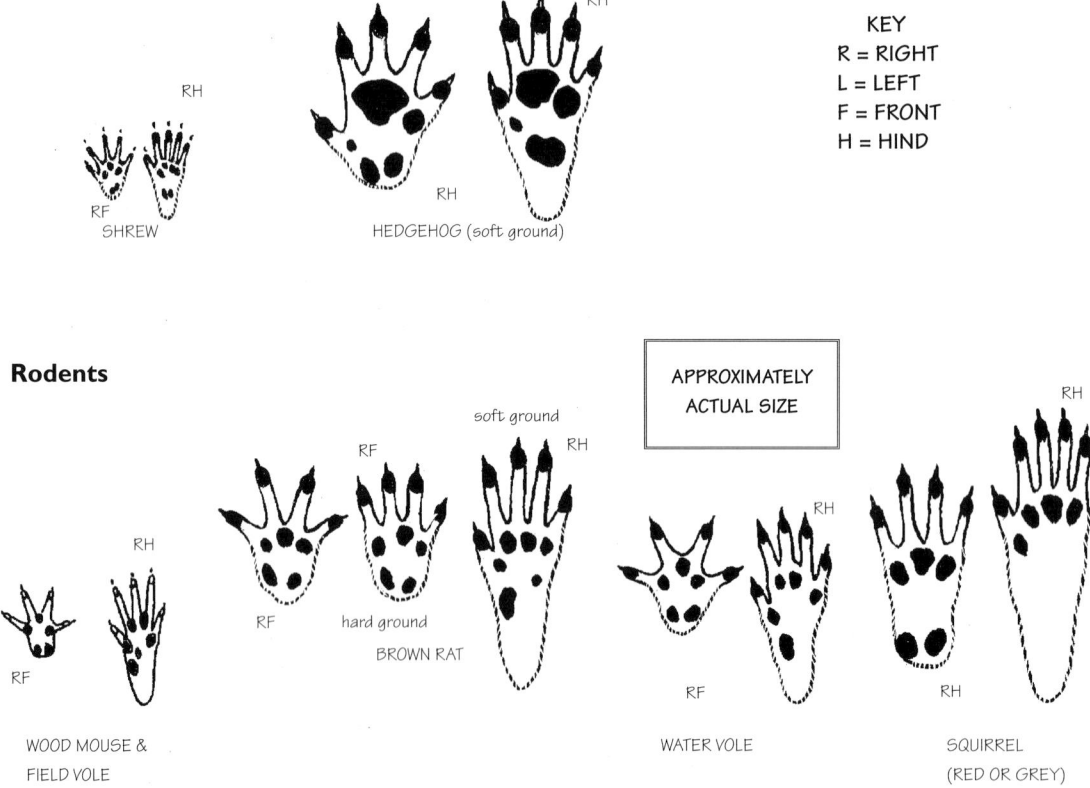

KEY
R = RIGHT
L = LEFT
F = FRONT
H = HIND

RH

RF
SHREW

RH

RH
HEDGEHOG (soft ground)

Rodents

APPROXIMATELY
ACTUAL SIZE

RH

RF

RF   hard ground
BROWN RAT

soft ground
RF   RH

RH

RF
WATER VOLE

RH

RH

RH
SQUIRREL
(RED OR GREY)

WOOD MOUSE &
FIELD VOLE

RF

Rabbits and hares

The position of the feet is the most distinctive feature of rabbit and hare tracks, with the long hind feet parallel and the fore feet between them.

Direction of travel

FOREFEET

RF

RF

HARE TRACKS

HARE (soft ground)

RABBIT (soft ground)

Carnivores
Fox, dog and cat

Foxes and cats leave a similar pattern of tracks in a straight line with one foot placed directly in front of another. The prints can be distinguished by the absence of claws in cats. Dog prints vary in size according to the breed but the shape is consistently different from foxes. Dog prints are as broad as they are long whereas fox prints are longer and narrower; diamond shaped. The plantar pad on the fox's hind foot, is only a little larger than the toe prints. In dogs it is much larger. The tracks of a fox usually shows a purposeful path.

SCALE 1:1

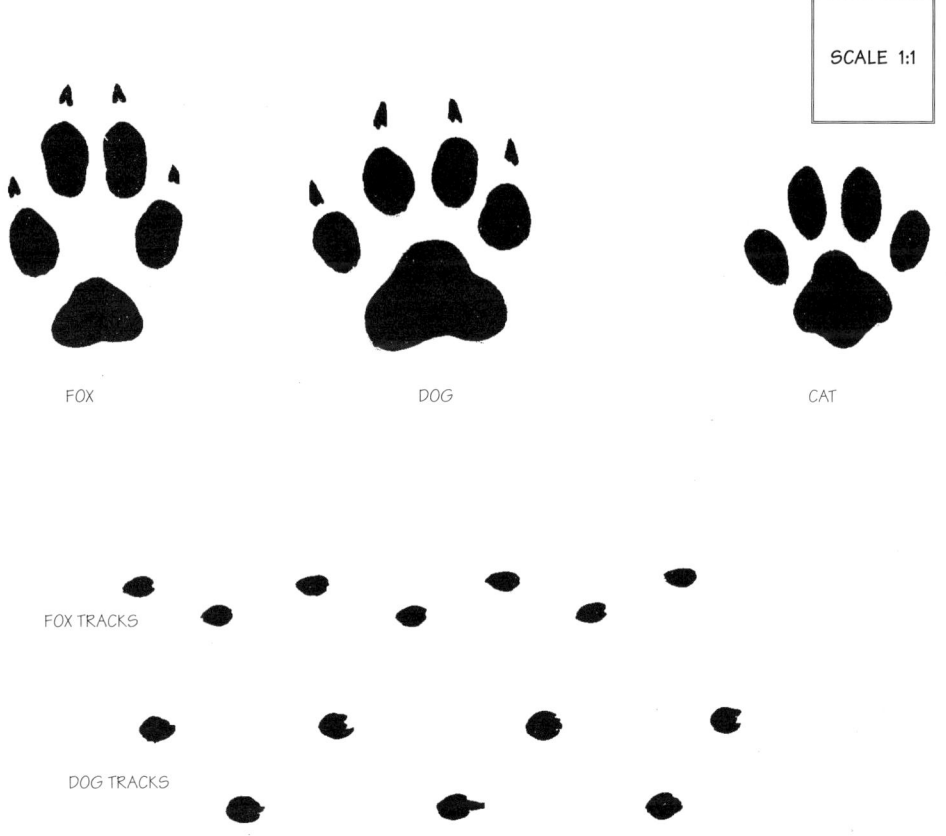

FOX

DOG

CAT

FOX TRACKS

DOG TRACKS

Mustelids

Mustelids tend to use a lolloping gait, due to their long back, except weasels which appear to run more evenly and badgers which amble. All leave a trail showing groups of three or four foot prints. All have a print showing five toes around a central pad although sometimes only four register.

OTTER MAKING TRACKS

RF

PINE MARTEN

RF

STOAT

RF

WEASEL

RF

POLECAT

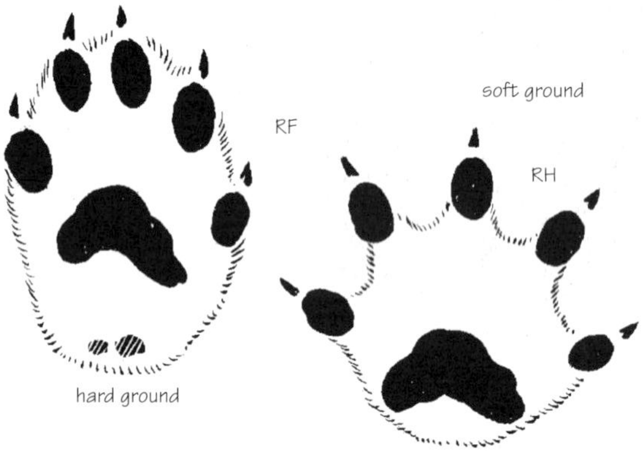

RF

hard ground

soft ground

RH

MINK

LH

BADGER

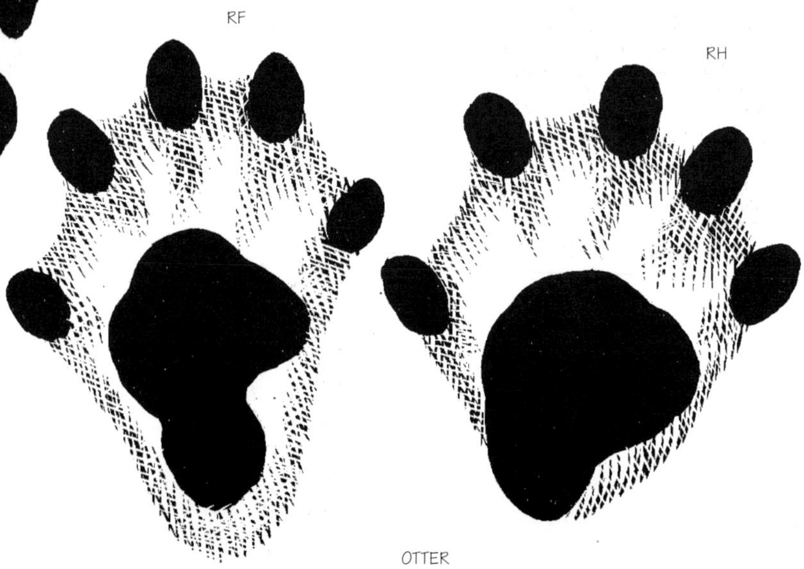

RF

RH

OTTER

APPROXIMATE ACTUAL SIZE

Ungulates (deer, sheep and goats) and pig

Deer prints vary in size with species but are fairly similar in shape. The shape, however, can vary according to how fast the animal is moving. When walking, deer tracks show two parallel bars but these become more splayed the faster the animal moves.

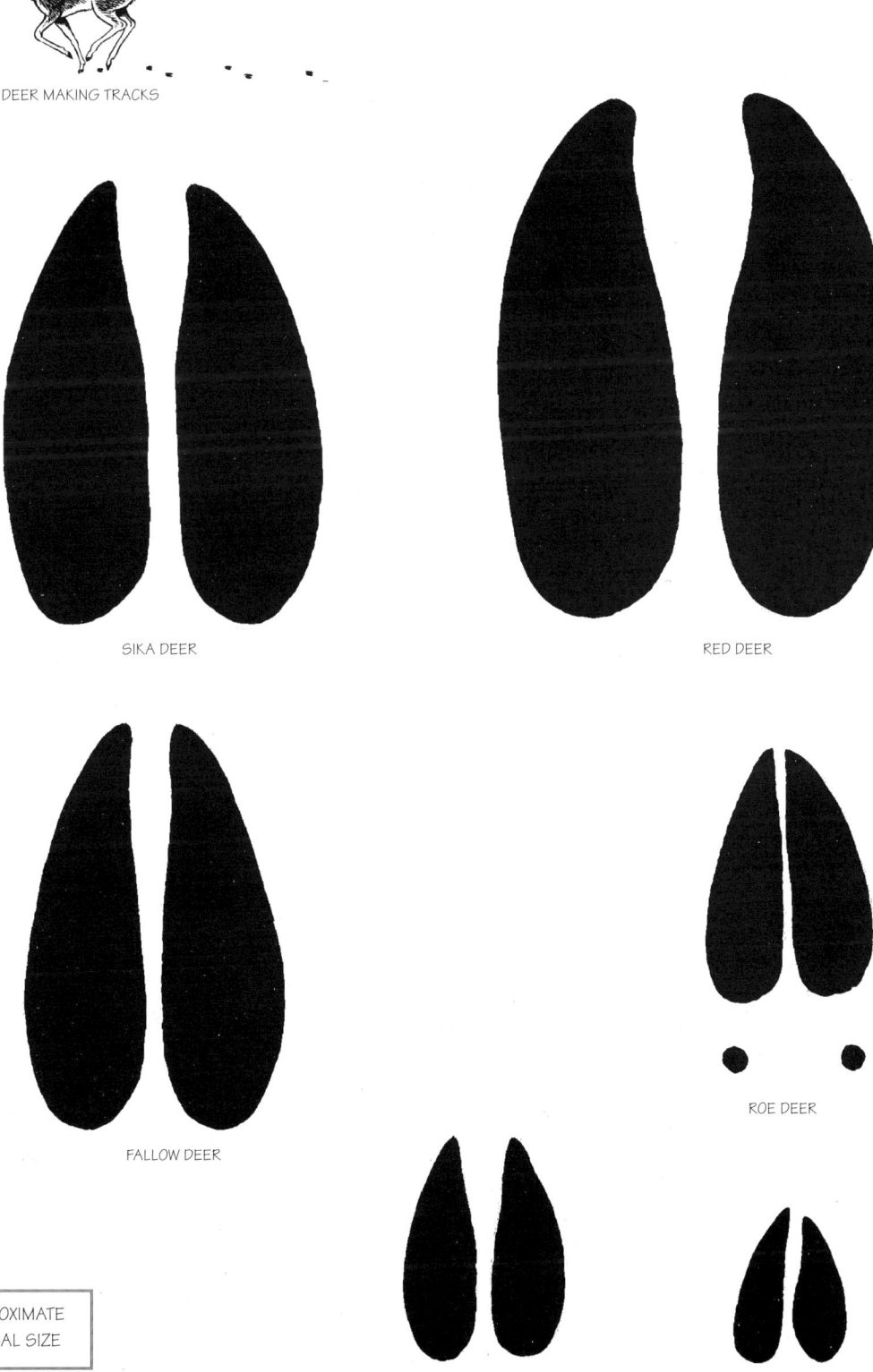

DEER MAKING TRACKS

SIKA DEER

RED DEER

FALLOW DEER

ROE DEER

APPROXIMATE
ACTUAL SIZE

CHINESE WATER DEER

MUNTJAC

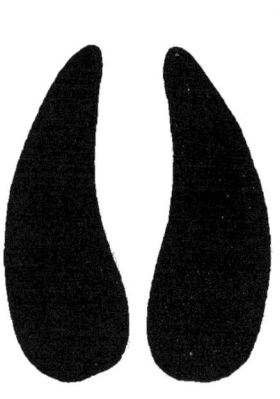

SHEEP

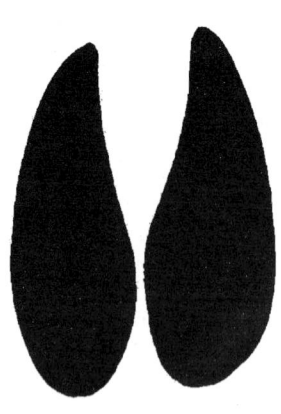

GOAT

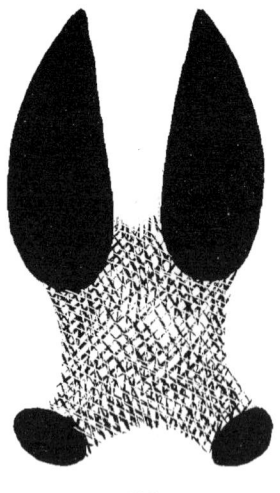

PIG

APPROXIMATE
ACTUAL SIZE

Trails

Trails form where the passage of an animal has caused disturbance to the substrate and vegetation. Badgers tend to form the most prominent trails which often lead from their setts to feeding grounds. Persistent use erodes and compacts the soil and slows the growth of grass on the trail, although vegetation may close over the top, forming a tunnel. Temporary trails can also be formed in early morning dew by passing mammals.

Investigate any linear features such as worn-looking paths. Paths may lead under fences and reveal further signs such as droppings, or fur caught on barbed wire or brambles. If the path leads to damp ground, perhaps by a river, stream or pond, it may be possible to determine tracks in the mud. Trails may also lead you to the animal's burrow or nest site and enable you to identify the species.

Droppings and latrines

Some mammals identify and communicate with each other by scent and leave their droppings in particular locations, and these may have a distinctive scent. Other mammals do not use their droppings for communication and deposit them at random.

N.B. If you intend to handle droppings wear surgical gloves, especially if you have cuts or grazes.

The droppings of insectivores are usually pellet- or sausage-shaped, and have a tendency to crumble to fragments of insect remains. Herbivores eat large quantities of vegetation and produce large amounts of faeces. These tend to be spherical or pellet-shaped and clustered, and have a uniform texture. Carnivores tend to produce single, larger, sausage-shaped droppings.

All British carnivores are mustelids, except for the fox and the wildcat. Musk glands at the base of the tail are particularly well developed in mustelids, and fresh droppings, which usually have hair and fur inside, smell of musk.

TABLE 4 indicates droppings which can be identified to species relatively easily. Small mammal droppings (shorter than 1 cm) are very difficult to identify to species and are therefore not described in detail. An exception to this is the droppings of water shrews which are less than 0.5cm in length and are the only small mammal to produce droppings containing fragments of aquatic insects. Variations in an animal's diet, digestion and 'extrusion dynamics' lead to distinct variations in content, colour, texture, smell, size and shape of droppings.

Droppings - shorter than 1 cm

Species	Mouse	Bat
Smell	House mouse - Strong urine Other mice - not so strong or no smell	Usually little or no smell
Texture	Very hard when dry - doesn't crumble	When dry crumbles to dust (insect remains)

APPROXIMATE ACTUAL SIZE

MOUSE

BAT

Droppings - about 1 cm or longer
Herbivores

Species	Water vole	Rat	Squirrel
Smell	Odourless	Foul, rancid	Depends upon diet but often sweet, hint of pine sawdust
Colour	Variable - dependent on food eaten. Usually dark green when broken up	Blackish Brown	Dark grey/black (Harris) Yellowish (Strachan)

WATER VOLE

RAT

SQUIRREL

Species	Rabbit	Hare
Smell	Sweet, damp digestive biscuit / hint of mown hay	Sweet, damp digestive biscuit / hint of mown hay
Colour	Yellowish brown-green	Greenish-brown
Distinctive feature	May be found in dense aggregations of pellets on prominent feature of habitat e.g. anthill	Larger and more flattened than rabbit, however, variable and dependent on diet

RABBIT

HARE

Species	Red	Fallow	Sika
No obvious smell or colour variation			

Species	Roe	Muntjac	Chinese Water Deer
No obvious smell or colour variation			

Insectivores

Species	Hedgehog
Smell	Sweet, hint of linseed oil
Colour	Blue-black
Distinctive feature	Crinkly, often studded with shiny fragments of insect cuticle

HEDGEHOG

APPROXIMATE ACTUAL SIZE

Carnivores

Species	Pine marten	Stoat	Weasel
Smell	Sweet, parma violets when fresh	Musky, but not too unpleasant	Musky
Colour	Blackish	Blackish brown	Brown

Features	Polecat	Mink
Smell	Foul, foetid meat, distinctly unpleasant	Foul, burnt rubber, rotten meat, unpleasant
Colour	Blackish	Greenish, black, brown
Distinctive feature	Deposited in prominent places	More bones than polecat but often found at similar sites and along river banks

Features	Badger	Otter
Smell	Foul, strong musk, oily, hint of hay	Sweet, jasmine tea, laurel flowers, slightly oily rag, distinctive. Can have a tarry smell. Often in very small quantities.
Colour	Blue-black-brown	Greenish, black-grey
Distinctive feature	Often containing seeds and large bits inside and often located in a pit or 'latrine'. A purple colour may be caused by blackberries in the diet. A formless dung-like mud is very common when large numbers of earthworms have been eaten.	Contains mainly fish scales, bones, shells of crustacae, feathers or fur.

APPROXIMATE
ACTUAL SIZE

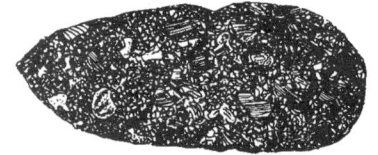

BADGER LATRINE -DROPPINGS ARE SLOPPY AND SHAPELESS

OTTER SPRAINT MARK - DROPPINGS ARE OFTEN SMEARED ON TO A STONE OR LOG

Feeding signs

Nibbled stalks of grass

Nibbled stalks of grass in small piles among tunnels in long grass indicate the presence of field voles. Where these are found along a river bank or canal they may indicate the presence of water voles and may incorporate a latrine (see DROPPINGS AND LATRINES for illustration).

FIELD VOLE FEEDING REMAINS

Gnawed hazel nuts

These are likely to be the food remains of bank vole, wood mouse, hazel dormouse or squirrel. The species can be identified by the gnawed marks on the nut. Squirrels split the nut leaving jagged edges and irregular pieces whereas the other three species gnaw a hole in a characteristic pattern. It has been suggested that holes gnawed by hazel dormice usually overlap with the scar on the nut.

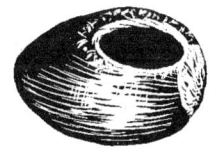

HAZEL DORMOUSE
Scratchy gnaw marks on outer edge of hole with
smooth, scooped out inner ring

Adult marks

Juvenile marks

SQUIRREL
Prized apart through
hole in top of nut

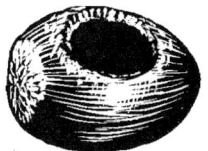

WOOD MOUSE
Toothmarks on outside of nut
and inner edge of hole

BANK VOLE
No marks on shell surface. chiselled
toothmarks on inner edge of hole

Burrows

Burrows are sometimes easiest to locate during the breeding season when activity is high and foraging parents produce well worn thoroughfares. The young may emerge but are unlikely to stray far. At other times of year, when the vegetation is low, it may be easier to spot burrows although there may be less activity.

Burrows of large mammals are generally located beside some kind of shelter such as a hedge or bank that gives the emerging animals protection. Outside the breeding season some mammals change burrow sites quite regularly. Wood mice, stoats, weasels and foxes sometimes appropriate a nest or burrow built by another species.

TABLE 5 indicates the approximate relative sizes of burrows of different species; it is meant as a guide, as sometimes individuals may dig particularly large or small burrows that fall outside the range indicated.

Table 5 **Identification of burrows**

Size	Diameter	How to remember size	Likely species
Small	Less than 3cm	Ping pong ball sized, 50p coin.	Small mammals i.e. mice and voles
Small to Medium	4-8cm	Tennis ball size	Mole, water vole, rat
Medium to large	8-20cm	Fist sized or larger	Rabbit
Large	More than 20cm	Football sized	Fox or badger

Small burrows - less than 4cm diameter

If food (seeds and berries) has been collected near the entrance of a small burrow, it probably belongs to a woodmouse or bank vole (Tattersall *pers. comm.*). House mice, which are generally restricted to buildings, can form distinctive holes in the wall and floor, creating greasy smears around the entrance. They often leave lots of gnawed food remains near the entrance and may leave a characteristic acetamide musty smell of urine. Longworth traps set at the burrow entrance can be used to catch and identify mammals in occupation. If you have access to a microscope you could try positioning sticky tape above the burrow, though not obstructing it, to collect hairs from animals passing in and out of the burrow. Alternatively a tracker board can be used.

In dense grass field voles often create a maze of runways just above ground level. Burrows are often made under corrugated iron sheeting left lying flat on the ground (see CORRUGATED IRON).

Small to medium to burrows - 4 to 8cm diameter
Mole
The characteristic spoil heaps of mole hills are familiar to most people. Mole hills consist of pure loose soil. Ant hills are usually firmer, consolidated by vegetation and roots - and also have ants in them!

Water vole and brown rats
Water voles and brown rats create burrows of similar size, within about 1 metre of river banks. Breeding water voles have a 'lawn' of nibbled grass in front of the entrance due to their habit of grazing in close proximity to their burrow. Brown rats, however, tend to have a fan of excavated soil in front of the entrance holes and are joined by well worn 'rat runs' which may even pass under water. In the breeding season water voles also have several 'latrines' within their territory, which have a pile of dark green droppings, some of which may have been trodden into a mush. Rats generally make their burrows near a food source and may also leave droppings nearby.

WATER VOLE BURROW

RAT BURROW

Medium burrows - 8 to 20cm diameter
Rabbit

Rabbit burrows are especially prevalent on slopes where drainage is more efficient. The entrance diameter can vary and may be up to 50cm.

Large burrows - more than 20cm diameter
Fox and badger

Badger setts account for nearly 60% of records of badgers held by BRC. They are particularly abundant on slopes on sandy soils. Fox 'earths' (which account for less than 2% of BRC fox records) tend to develop a 'foxy' smell and are generally a fairly messy affair. When the cubs are emerging the surrounding vegetation may be flattened, and feathers, chewed bits of wood and bones may be strewn about the entrance. In contrast, badgers do not usually leave food debris lying about, although there is often a large heap of bedding and dry grass thrown out in front of the entrance to the sett. This often contains badger hair which is distinctive. These hairs are long, wiry and white with a broad black zone towards one end and tend to be slightly wavy, not dead straight (See the HAIR section). Tracks in soil surrounding the entrance may also indicate the species of the user.

Badger setts tend to have more than two entrances. Foxes, which are not equipped with such powerful digging apparatus as badgers, dig smaller and shallower burrows, tend not to create more than two openings and have a smaller spoil heap at the entrance. To confuse the matter, foxes often take over old badger setts.

FOX EARTH

BADGER SETT

Otter and mink

Otters make holts in cavities under roots of bank side trees or boulders, but may also rest in couches of flattened vegetation or scraped out soil, often concealed by surrounding vegetation. Mink will also make dens in the hollows of bankside trees.

OTTER HOLT

OTTER COUCH

Other types of hole

Dormice, squirrels and pine martens make use of holes in trees. However, a hole in a tree cannot be taken as a reliable record. If you find a hole you believe is used by one of these species, try to find other signs to back up the record.

How to Find and Identify Mammals
The Mammal Society

39

Nests

Mammals which build nests above ground include hedgehogs, harvest mice, dormice, and squirrels.

Red and grey squirrel

The nests of red and grey squirrels, known as dreys, are spherical collections of twigs and leaves which are generally (but not always) located in a fork in the branches close to the trunk, usually from 6m up. It is not generally possible to distinguish between the dreys of red and grey squirrels. In contrast, the nests of magpies, crows and other birds which build nests of similar size can be distinguished by the fact that they tend not to incorporate leaves and are usually located further from the tree trunk.

SQUIRREL DREY
WITH LEAVES ATTACHED TO TWIGS

CROW OR MAGPIE NEST MADE OF DEAD
TWIGS SO NO ATTACHED LEAVES

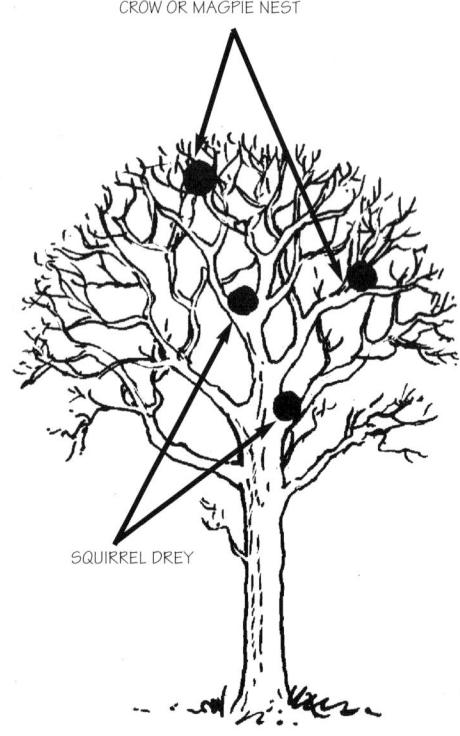

CROW OR MAGPIE NEST

SQUIRREL DREY

POSITION OF SQUIRREL AND CROW OR MAGPIE
NEST IN TREE, ALTHOUGH SQUIRRELS
SUMMER DREYS CAN BE POSITIONED
FURTHER AWAY FROM THE MAIN TRUNK

Table 6 **Identification of some distinctive small mammal nests**

	Harvest mouse	**Hazel dormouse**	**Hedgehog**
When to find nests	Easiest in early autumn	Summer/winter	Easiest in late winter
Where to look for nests	Farmland, hedgerows, patches of tall stiff-stemmed grasses and reeds	Coppiced hazel woodlands especially in bramble	In sheltered cavities e.g. between logs in hedgerows, gardens, farmland and under low brambles
Height above ground	Depends on height of the supporting plants (see diagram overleaf) Up to 1.5m	1- 2m off the ground especially in bramble	On the ground
Size of nest	Golfball to cricket ball-sized	Grapefruit-sized	50cm diameter
Distinguishing features of nest	Composed of leaves of grasses shredded longitudinally, some of which are still attached to the stems. The nest, therefore, appears to hang from the surrounding stems. The nests of reed warblers, however, have stems of supporting plants passing through the nest. They are also domed, not cup-shaped like birds nests in grass.	May contain woody species woven into the nest, such as strips of honeysuckle bark with an outer layer of leaves.	Composed of leaves, grass and other materials. The leaves are gathered into a heap and the hedgehog then shuffles around inside until the leaves become similarly orientated and packed flat against each other.

Harvest mouse

The harvest mouse is an under-recorded species. In summer, harvest mice spend most of their active hours climbing in long grasses and tend not to be caught in Longworth traps, which are usually set on the ground. The incidence of harvest mice in owl pellets is low compared to other small mammals but appears to be rising as, perhaps, other prey species decrease in abundance. The characteristic breeding nests, however, provide obvious signs of the presence of harvest mice and are unlikely to be confused with other types of nest. Looking for nests requires less effort than trapping the animals themselves and allows a large number of sites to be surveyed in one day.

HARVEST MOUSE AND ITS NEST

POSITION OF HARVEST MOUSE NESTS IN DIFFERENT VEGETATION TYPES:

IN REEDS (e.g.Phalaris),
LONG GRASSES AND CEREALS

IN PLANTS THAT FORM TUSSOCKS OF LEAVES AT THE BASE
OF THE FLOWERING STEMS
(e.g. cocks foot (Dactylis glomerata)
tuffed hair grass (Deschampsia caespitosa) and Molinia spp.)

Hazel dormouse

Dormice spend most of their time in trees and overgrown hedgerows and are rarely caught in Longworth traps. Apart from using their own nests (see Table 6) dormice may use a deserted bird nest or nest box.

HAZEL DORMOUSE SUMMER NEST

Hazel dormouse hibernation nests are more difficult to locate. They are built on the ground, usually under moss or leaves. They are tightly woven, tennis ball-sized and made of shredded fibrous material (e.g. honeysuckle bark).

Fat dormouse

The fat dormouse often uses deserted squirrel dreys, tree holes, cavities in roof spaces or large bird nest boxes in summer and, in winter, hibernates underground.

Hedgehog hibernation nests

Hedgehogs may build several nests in the course of one winter. These tend to be tucked under a bush, log pile, garden shed or anywhere that offers support and protection.

HEDGEHOG WINTER NEST

Corrugated iron sheets

Small mammals are often attracted to the warm, sheltered conditions generated underneath a metal sheet. The sheet can then be lifted quickly to catch any small mammals sheltering beneath it. This method can be used to improve the chance of recording field voles, bank voles, yellow-necked mice and shrews.

Tins and boards may be inspected several times during the season and it may be worth noting where they are on a map. Do not place these sheets on other people's land where they may cause problems, without asking permission, but make use of any that are already in place.

Size of sheet

The more sheets you inspect the more likely you will get a positive record. The main influence may be the size you can carry to the site and turn over quickly when checking. A sheet of around 0.5m x 1.5m is ideal.

How to position the sheet

Get permission from the landowner first. Position the sheet where the sunlight can warm it, but not too much or the sheet will become too hot. Try to locate it near some cover to provide a mixture of sun and partial shade and to prevent temperature extremes. Try to position the sheet on grass. Before checking for mammals wait until the grass has died back underneath.

How to turn over the sheet - hints on handling

Two people should be present, one to lift the sheet and one, wearing gloves to avoid being bitten, to swiftly grab any small mammals underneath. The person catching the animals may find it easier to gently pin the animal down with the palm of one hand and then to get the scruff of the neck with the other hand. Alternatively, place a piece of cloth over the animal before getting it by the scruff of the neck. Shrews and voles can be held by the tail. Mice should never be handled in this way (except harvest mice) as the skin strips off the tail easily.

Be aware that adders and other reptiles also favour the conditions found under metal sheets in sunny positions.

Case study 3 **Recording small mammals under corrugated iron**

> Caught my first yellow-necked mouse under a sheet of corrugated iron!
>
> *Pat Morris*

Bait stations

Water shrews, in common with other shrews, are inquisitive and will readily investigate novel objects, particularly if a food source is detected. Bait stations are composed of a length of plastic piping (20cm long; 4cm diameter) with muslin netting held over one end by an elastic band. By providing a suitable food source (Blowfly *Calliphora* pupae), these tubes encourage visiting shrews (and other small mammals) to enter, linger to feed and defaecate. The droppings of water shrews, which have been feeding in aquatic habitats, can be distinguished from other small mammals by the presence of undigested remains of aquatic prey. Aquatic prey can be distinguished from terrestrial invertebrate prey items by the distinctive shapes, frequency of hairs and occurrence of particular structures such as spines. For illustrations of the characteristic remains of aquatic insects in water shrew droppings see Appendix 5.

How to Find and Identify Mammals
The Mammal Society

Mammal identification from hairs

The coat of most mammals is composed of a layer of fluffy insulating hair, 'underfur', with a thin layer of stiffer, longer guard hairs on the outside and particularly on the back of the animal. The guard hairs are the easiest to identify and are often caught as the animal passes and presses against something sharp such as barbed wire fencing or brambles, or something sticky. Identification, however, is not always easy and it can sometimes be hard to match what you can see through the microscope with the pictures in the book. Only intact hairs should be used for identification. Do not use parts that have been damaged by a sticky surface.

BADGER HAIR
(to scale)

FUR CAUGHT ON BARBED WIRE

Guard hairs can be identified according to their colour, length, form and structure. The cross section of guard hairs is often characteristic; shrew hairs, for example, have grooved surfaces. The scales on the outer surface of the hairs are also distinctive but are hard to see under a light microscope without special preparation.

Preparation for examining small hairs under a microscope

1. Dissolve a few grains of gelatin in half a small egg cup of water, paint a thin layer on to a microscope slide and leave to set for ten minutes.

2. Select a few stiff guard hairs. Do not use fine or crinkly ones.

3. Place a few hairs on the gelatin film and leave for 30 minutes to set.

4. Peel off the hairs using tweezers and examine the impression of the surface scales under the microscope at x 50 to 100 magnification. The most characteristic part of the hair for identification purposes is the 'shield'; the mid to upper third of the hair.

Alternatively try using clear nail varnish film on a microscope slide instead of gelatine.

Hair tubes

Hair tubes are lengths of tubing (generally plastic) about 10cm long that are positioned in sites where they are likely to be used as a tunnel by mammals of an appropriate size. The 'roof' inside the tube is lined with sticky tape, such as carpet tape, to trap hairs from passing mammals and which can be removed for identification. The diameter of the tube should be determined by the type and size of mammal you are attempting to record, although it is possible to raise the level of the 'floor' by pouring in and setting Plaster of Paris, or glueing in a rigid strip of waterproof material. Seeds in a matrix of peanut butter is an attractive bait to use for small mammals and can be smeared on the inside of the tube.

Table 7   **Hair tube diameters**

Species	Hair tube diameter in cm
Common shrew	3.0
Pygmy shrew	2.5
Water shrew	3.5
Bank vole	3.5 - 4.0
Field vole	3.5 - 4.0
Wood mouse	3.5
Yellow-necked mouse	4.0
Harvest mouse	3.0
House mouse	3.5
Dormouse	3.5

Part 5 Recording dead mammals and parts of mammals

Road casualties

Equipment

1. General recording form and, if possible, a dictaphone.

2. Clip recording form on to dashboard or keep in glove compartment with a pencil so you can easily stop and make a note, or your passenger can do this.

3. Ordnance Survey maps

4. Bin liners for those likely to find polecats or polecat-ferrets (these should then be sent for confirmation to Andrew Kitchener, Dept. of Natural History, Royal Museum of Scotland, Chambers Street, Edinburgh EH1 1JF).

5. Small plastic bags to use as gloves and to collect hair which, in difficult cases, you may need to send to your county mammal recorder for identification or confirmation.

6. Sellotape to stick hair to a sheet of paper with details of the record, and envelopes to put it in.

Examining mammals killed on roads allows us to keep records of some larger, more elusive species in the community, and sometimes confirms the presence of rare species. This method probably under-samples small mammals as their carcasses are harder to see from a moving car and are more likely to be carried off whole by scavengers. TABLE 4 indicates mammals most likely to be recorded killed on roads. This is the main method of gathering records for some species (e.g. hedgehog, pine marten) and a very important one for others (e.g. stoat, otter, wildcat and brown hare).

Early morning is the best time to collect records of road kills, as mammals tend to be knocked down at night when they are likely to be confused by oncoming headlights and when they are also most active. Mammals squashed overnight will be easier to inspect early in the day before further traffic squashes them even more.

It is often possible to reliably identify a large carcass at 40mph. A dictaphone is useful for recording the details although you should stop the car to do this. If you can't identify the carcass, stop for a closer look from the verge. If you need verification collect some hair, attach it to sellotape and send to your County Mammal Recorder, for possible identification.

Case study 4 Recording mammal road kills

In the summer of 1993 the body of an otter on a road 20 km from London confirmed the return of this species to the River Thames.

Rob Strachan

Leaving the car to retrieve carcasses

Safety for yourself and other road users is the paramount consideration. Traffic on busier roads tends to result in more squashed mammals. To prevent becoming one yourself follow these guidelines.

Never collect on motorways as it is illegal to stop on the hard shoulder except in an emergency. Some busy roads often have parking laybys at frequent intervals which are worth using. If there are no laybys try to return to the site and park about 30m ahead of the carcass using the car as a shield, pulling onto the extreme nearside. Put on your hazard warning lights and, subject to traffic flow, leave the car from the verge side. Put on something fluorescent if possible and walk up the verge rather than the kerb. When the road is clear retrieve the carcass quickly by lifting it on to a bin liner and carrying it back to the car.

It may be worth contacting your local motorway services and county council department responsible for retrieving carcasses from roads, and supply them with mammal recording forms to return to the County Mammal Recorder.

Discarded bottles

Equipment

1. Rubber gloves or clear plastic bags to use as gloves
2. Water
3. White dish, plate or piece of paper on which to tip contents of bottle for inspection
4. Small mammal skull identification key

Over 100,000,000 bottles a year are lost from circulation. Apart from contributing to Britain's litter problem, many of these bottles, depending on where and how positioned, trap small mammals. Although most bottles that have trapped mammals contain only one or two, a milk bottle in Essex was found containing 28 small mammals. About 5% of bottles containing mammals have killed more than 7 (*Morris, 1970*).

Mammals probably enter out of curiosity or in search of food. Entering may be comparatively easy especially if the bottle neck is pointing up, however, escape is harder as the animal must push against slippery glass which may be wet, and against a narrowing funnel. Animals probably perish from cold or starvation. Prevent further deaths by collecting and disposing of bottles if possible, or positioning them with the neck stuck into the soil and pointing downhill, but it is worth first looking at the contents and extracting any information you can.

As bottles have sometimes been *in situ* for extended periods they sometimes account for records of species which are rare and not commonly found by other methods (e.g. water shrew). Bottles provide a useful starting point when drawing up a species list for a site (e.g. a nature reserve) and are also useful for sampling inaccessible habitats where the repeated visits required by Longworth trapping are inconvenient. In general common shrews are the most common captures and, along with the wood mouse and bank vole, account for over 90% of all mammals in bottles.

The best season for bottle surveys is late winter when the vegetation is low and bottles are easiest to find. Tip out the contents of the bottle and wash away material other than bones. Pick out the skulls and identify them using the key given in APPENDIX 4. Pay particular attention to the upper jaw and then remove teeth to look at tooth sockets as the key suggests.

Case study 5   **Recording mammals in discarded bottles**

Bottles are highly effective, lethal mammal traps. By collecting discarded bottles that lie strewn about the countryside, you can save many small mammals from an unpleasant end and gather useful records at the same time. The altitude record for the pygmy shrew, for example, came from a bottle near the peak of Ben Nevis. There can be further bonuses - I once found an unopened bottle of champagne in a ditch! Do not, however, put out bottles specially to catch mammals.

Pat Morris

Owl pellets

Equipment

1. Tray of warm soapy water
2. Binocular microscope or hand lens
3. Small mammal skull identification key
4. Owl pellet recording form

Owls are one of the most effective hunters of small mammals. They swallow their prey whole, digest the flesh and then deposit the undigested remains in neat pellets of bones and fur. Analysis of the contents of pellets allows the small mammal community to be sampled (with a bias towards mammals which are favoured prey) and can be particularly useful for sampling less common species which are unlikely to be found in the course of small-scale trapping efforts.

Barn owl pellets are the easiest to study as barn owls tend to use sheltered roosts, such as outbuildings, where the pellets are kept dry and intact. They also tend to be faithful to their roosts so many pellets accumulate in one place. Tawny owls may use a number of roosts which tend to be in exposed locations such as hollow trees, therefore, their pellets are often widely scattered and soon disappear. Herons, gulls, crows and birds of prey also take mammals but tend to tear their victims apart, often leaving the head (the most important identifying feature) and sometimes digesting the bones, therefore their pellets form a less complete record of their diet.

A six figure grid reference of the owl pellet source should be recorded although a four figure grid reference should be used to approximate the location where the prey was captured on distribution maps of the prey. It is not possible to assess the age of owl pellets, therefore, if you want to sample the contemporary mammal community collect the pellets on a regular (e.g. monthly) basis.

Fox droppings may be mistaken for owl pellets. Look for the pointed ends on fox droppings compared with owl pellets which have rounded ends. Foxes also tend to chew their prey and break up the bones. Fox droppings often smell strongly whereas owl pellets do not.

OWL PELLET WITH MAMMAL REMAINS FROM A CLEANED PELLET

Pellet analysis

It is easiest to concentrate on one pellet at a time. Either tease the pellet apart using tweezers or soak the pellet in warm, soapy water to soften the matted hair and dispel insect larvae. Separate out any skulls first as these are the easiest bits to identify, followed by the larger bones. Place the skulls and bones on a dark background and compare the skulls and then the teeth from the lower jaw with the key in APPENDIX 4. The identification process may be slow to start with but you will quickly learn the limited options.

Alternatively, pellets may be sent to The Mammal Society Owl Pellet Survey. This ongoing survey started in January 1993 and to date (March 2003) has records of over 103,700 prey items recovered from over 29,300 pellets received from 221 locations. 20 small mammal species have been identified, with field voles, wood mice and common shrews accounting for over 79% of the prey items.

Batches of owl pellets should be gathered at regular intervals and sent to:

Alistair Love, 4 Laurel Way, Totteridge, London N20 8HP Tel: 020 8445 8989

Cat kills

Cats can be a useful source of records depending on the hunting ability of the individual and on its hunting grounds. Any small mammals, up to the size of a young rabbit may be taken though common shrews, field voles, bank voles, wood mice and house mice are probably the most common captures. The drawback of using your cat as a sampling method is that it is hard to know exactly where the prey was caught. It is probably safest to give a four figure grid reference on the recording form with the six figure grid reference (where you retrieved the prey item) in brackets with an explanation.

Case study 6 Recorded mammal cat kills

> While I was living in Durham over a ten year period, our cat successfully caught and brought to me the third and fourth records of the water shrew in the county.
>
> *Rob Strachan (1995)*

Parts of mammals

Considering how many wild mammals die each year it may seem surprising that we come across so few carcasses. The vast majority of wild mammals either hide themselves away to die and decay in nests or burrows, or, are eaten by carnivores and digested to an unrecognisable extent. Remains are also carried off by scavengers. Occasionally, however, a carcass is encounted where one needs to identify the remains. Where the carcass has decayed leaving only the skeleton for identification examine the teeth, skull and limbs.

Teeth

Prominent canines indicate the individual is a carnivore; these are used to grip the prey. The absence of canines with prominent incisors suggests the individual is a herbivore. The incisors are used to bite vegetation from the plant and may be replaced in the upper jaw by a horny plate in deer, cattle and sheep. There may be a large gap between the incisors and the molars.

Herbivore teeth have crescents on their grinding surfaces; carnivore teeth are pointed unless worn smooth. Muntjac and male Chinese water deer, however, develop large canines or 'tusks' for fighting their rivals.

The skull

1. **The position of the eye sockets**
   Carnivores have forward pointing eyes to allow them to assess the distance of their prey effectively. Herbivores have eyes on the sides of the skull to allow predators to be detected by a wider field of view.

2. **The attachment of jaw muscles**
   It is possible to identify a mature badger skull by a distinctive sagittal crest running along the middle of the skull. This ridge provides attachment for the powerful jaw muscles. In adult badgers the lower jaw does not detach easily from the skull.

Antlers

Cast antlers can be identified from their shape (see DEER under MAKING SIGHTINGS). Cast antlers have often been chewed or nibbled by rodents.

Part 6 Live trapping

An obvious way of sampling the small mammals in an area is to catch and identify some of them. Small mammal trapping can be organised relatively easily but specialist licences must be obtained from the statutory agencies for trapping shrews, dormice and large mammals.

Small mammal trapping

Live-trapping small mammals is a fascinating way to learn more about them and gives you the opportunity to observe the animals in the hand. However, it needs to be done with care and patience and can be time consuming. If trapping is being carried out as part of an exercise to estimate the population size or home range of small mammals refer to The Mammal Society's more detailed texts where further tips on trapping may be found.

Longworth trapping

Equipment

1. Longworth traps (minimum of 20 for full survey of a site)
2. Hay to provide bedding in the trap
3. Seed for bait and to feed trapped rodents
4. Insectivorous food for shrews such as blowfly pupae or 'casters' (available from fishing shops), or alternatively, have a 13mm hole drilled in the side of the trap to allow shrews to escape
5. Flags (colour tags on bamboo canes) to mark the site of each trap
6. A clear polythene bag in which to empty the trap and check for captures

The standard and most widely used small mammal trap is the Longworth trap. This is composed of a tunnel, with a trip mechanism and trap door, and a nest box for food and bedding to keep captures alive. Longworth traps are expensive (about £35 from Penlon Ltd) although members of The Mammal Society can borrow traps from the trap loan scheme. Plastic 'Trip Traps' can also be obtained from Penlon Ltd or pet shops (make sure you get one with a 'rest box'). These traps are less robust and often less effective than Longworth traps but they are considerably cheaper.

If shrews are likely to be caught (i.e. if there is no hole made specially) or if you deliberately want to catch them, a licence should be obtained from the relevant statutory agency (English Nature, Countryside Council for Wales or Scottish Natural Heritage). Alternatively, contact The Mammal Society to obtain cover through The Mammal Society's block licence (members of The Mammal Society only).

Traps should be set in the evening so that captures are not incarcerated all day and all night and they should be visited at least twice a day; morning and evening, to check and release captures. With between 10 to 20 traps a range of species can be caught over a three day period. Spread the traps out (say 20m apart) and try to cover the range of habitats e.g.woodland, woodland edge, grassland and wet areas within a site. Mark the points to allow you to locate them again with a marked bamboo cane for example, or, to avoid drawing attention tie bundles of grass around overhanging branches.

Successful trapping depends on where and how the traps are set and the efficiency of the trapping mechanism of your particular traps. For a complete guide to Longworth trapping obtain The Mammal Society publication *Live Trapping Small Mammals - A Practical Guide (Gurnell and Flowerdew)*.

LONGWORTH TRAP (should be covered when set)

Pitfall trapping

Pitfall traps are cylinders such as jars or tins sunk into the ground until the top is level with the soil surface. Small animals fall into the trap and cannot escape up the smooth sides of the container. It is, therefore, essential that when the traps are not in use either they have a well fitting lid or are removed from the site. Shrews, field and bank voles can be trapped in jam jars. Mice require deeper traps which are generally impractical. Food should be placed in the trap and pieces of narrow board laid across the top to prevent deer catching their feet in them. Pitfall traps should be checked regularly (e.g. every few hours).

Nest boxes

Dormouse boxes

Dormouse boxes, similar to bird boxes but erected facing a tree trunk, are one of the few ways of studying dormice and, again, their success depends on how and where they are positioned. You do not need a licence to set up boxes but you do need one as soon as you find a dormouse inside, in order to examine it. You might also find wood mice, yellow-necked mice and other species in dormice boxes therefore it is advisable to wear gloves when checking the boxes. For further details obtain The Mammal Society publication: A PRACTICAL GUIDE TO DORMOUSE CONSERVATION (*Bright & Morris 1989*).

Dormouse nest tubes

These can be used in the same way as nest boxes to check for the presence of dormice. They consist of two parts, a wooden tray and plastic nesting tube. These tubes are much smaller and lighter than boxes and especially useful in areas on non-traditional habitat such as hedgerows or areas of scrub. Nest tubes can be purchased from The Mammal Society.

Case study 7   Recording small mammals in nest boxes

The two largest yellow-necked mice ever recorded, weighing over 50g, were both caught in nest boxes.

Pat Morris

Sending in mammal records

The recording forms (EXAMPLE FORMS 1-5) are designed to collect the information needed for effective biological recording in an easy-to-use format. The layout of the form also enables efficient data entry into a computer database.

Recording forms should be sent in to the County Mammal Recorder, ideally, on a quarterly basis or as they request. This allows him or her to manage the data more easily and to provide regular feedback, such as newsletters, to observers. To find your County Mammal Recorder you can contact The Mammal Society on 020 7350 2200 or email: enquiries@mammal.org.uk or see our website: www.mammal.org.uk

Summary

Mammal records can be collected in many ways. It requires patience, skill, knowledge and luck but this makes it all the more rewarding. This manual attempts to provide the mammal observer with the skill and knowledge needed to collect valuable mammal data which can be used for mammal conservation. The manual covers the range of methods which can be used, outlines the critical information needed to use each method and recommends further relevant sources of information where appropriate.

References and further reading

Arnold, H.(1994) ATLAS OF BRITISH MAMMALS Institute of Terrestrial Ecology HMSO. (out of print)

Bang, P. & Dahlstrom, P. (2001) ANIMAL TRACKS AND SIGNS. Oxford University Press, Oxford*

Bellamy, D. (principal consultant) (2000) THE COUNTRYSIDE DETECTIVE. The Reader's Digest Association Limited, London

Bright, P. & Morris, P. (1989) A PRACTICAL GUIDE TO DORMOUSE CONSERVATION. The Mammal Society, London*

Brown, R.W., Lawrence M.J. & Pope J. (1992) ANIMAL TRACKS, TRAILS AND SIGNS. Hamlyn, London. (out of print)

Corbet, G.B. & Harris S. (Eds.)(1991) THE HANDBOOK OF BRITISH MAMMALS Third Edition Blackwell Scientific Publications, Oxford. (out of print)

Evans, P. G.H. (1995) GUIDE TO THE IDENTIFICATION OF WHALES, DOLPHINS AND PORPOISES IN EUROPEAN SEAS Sea Watch Foundation Publication, Oxford.

Gurnell, J. & Flowerdew, J.R. (1995) LIVE TRAPPING SMALL MAMMALS - A PRACTICAL GUIDE The Mammal Society, London.*

MacDonald, D. & Barratt, P. (1993) MAMMALS OF BRITAIN AND EUROPE Collins Field Guide, Harper Collins, London.

Morris, P. (1970) THE STUDY OF SMALL MAMMAL REMAINS FROM DISCARDED BOTTLES School Natural Science Society Publication Nº· 41

Morris, P. A. (Principal consultant) (2001) ANIMALS IN BRITAIN Nature Lover's Library, Reader's Digest, London.

Morris, P. A. and Wroot, S. PREPARATION OF MAMMAL SKINS FOR SCIENTIFIC, EDUCATIONAL AND DISPLAY PURPOSES The Mammal Society, London. (out of print)

Strachan, R. (1995) MAMMAL DETECTIVE Whittet Books, Weybridge.*

Teerink, B. (1991) ATLAS AND IDENTIFICATION KEY TO HAIR OF WEST EUROPEAN MAMMALS Cambridge University Press, Cambridge. (out of print)

Yalden, D.W., (2003) THE ANALYSIS OF OWL PELLETS The Mammal Society, London.*

Laminated Guides

Bullion, S. (2001) A KEY TO BRITISH LAND MAMMALS The Mammal Society and Field Studies Council, London*

Bullion, S. (2001) A GUIDE TO BRITISH MAMMAL TRACKS AND SIGNS The Mammal Society and Field Studies Council, London*

Jones, K & Walsh, A. (2001) A GUIDE TO BRITISH BATS The Mammal Society and Field Studies Council, London*

* Books available to buy from The Mammal Society

The Mammal Society also produces a number of booklets on British Mammals ranging from Practical Guides to field techniques to The Mammal Society Series of books on individual mammal species.

For a copy of our Publications List or to buy any of these books contact The Mammal Society on 020 7350 2200 or www.mammal.org.uk

Appendix 1 **How to read a grid reference**

A map of Great Britain is covered by 100 km squares, each of which is identified by a set of two letters.

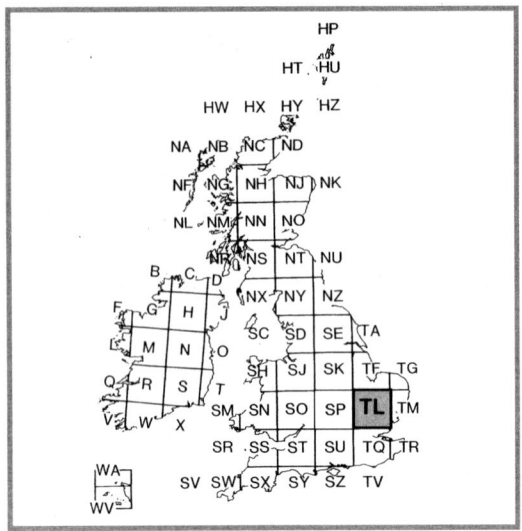

On Ordnance Survey maps these 100 km squares are further sub-divided into smaller squares by grid-lines representing 10 km spacing, each numbered from 0 to 9 in an easterly (left to right) and northerly (upwards) direction from the southwest corner.

Using this system you can identify the 1:25,000 Pathfinder maps of your area. For example, the shaded square here is TL 73.

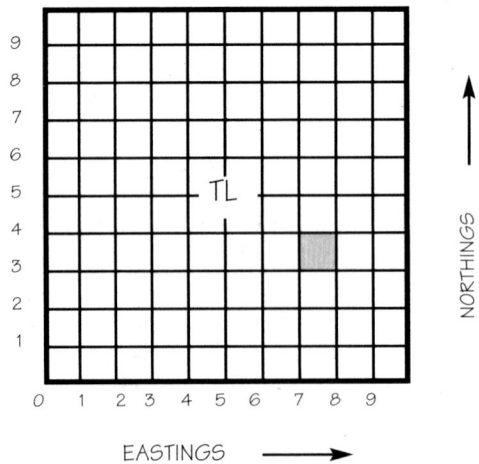

After the letters you quote the eastings first - then the northings. If you have trouble remembering, say "along the hall THEN up the stairs" or easting and then northing (in alphabetical order).

On Ordnance Survey Landranger maps, you can find the two grid letters on the legend or on the corners of the map. The grid has also been further divided into 1km intervals.

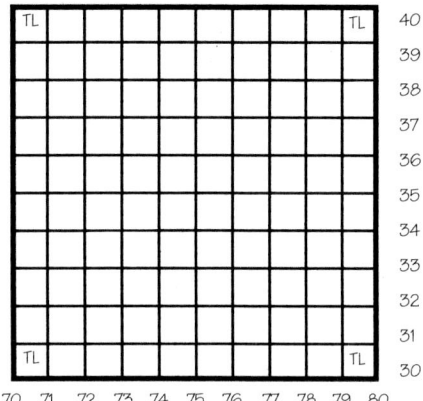

By estimating the eastings and northings to one tenth of the grid interval you can specify a full six figure map reference, accurate to within 100 metres on the ground. All you do is guess how many tenths away from the grid your point falls. Half way to the next grid is five tenths - and so on. Quote all the eastings first- then the northings.

The 100 metre grid reference is shown like this: **TL 763317**; an example of a 6 figure reference.

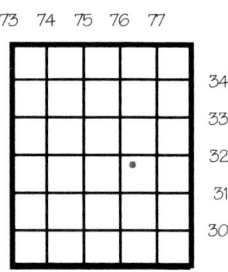

Here is a 1:50,000 Landranger map extract. Use the National Grid reference system to find:
The Church at **SK 568 970**

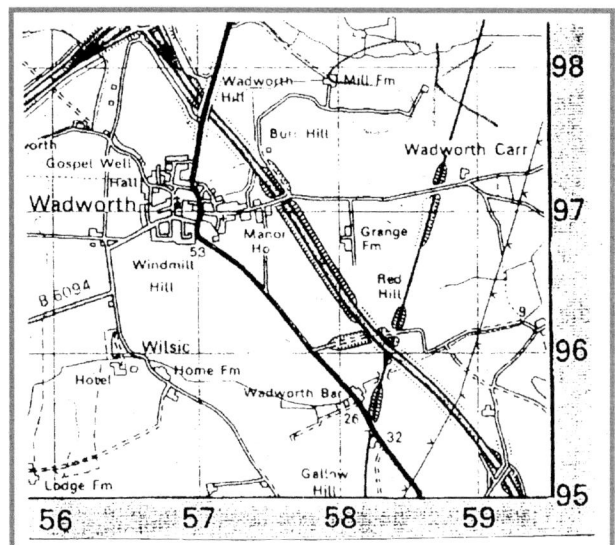

Modified from the
Ordnance Survey leaflet:
HOW TO TAKE A GRID REFERENCE

Appendix 2 Species names and commonly used synonyms

Species	Synonyms
Hedgehog	Urchin, Hedgepig
Common shrew	Shrew mouse, Ranny
Pygmy shrew	Lesser shrew
Water shrew	Otter shrew
Rabbit	Coney
Brown hare	Common hare
Mountain hare	Blue hare, arctic hare, Irish hare
Red squirrel	Common squirrel, Brown squirrel
Bank vole	Red-backed vole
Field vole	Short-tailed vole
Water vole	Water rat
Wood mouse	Long-tailed field mouse
Harvest mouse	Dwarf mouse, Red mouse
Brown rat	Common rat, Sewer rat, Norway rat
Hazel dormouse	Common dormouse
Polecat	Foul marten
Badger	Brock
Pine marten	Sweet Marten

Appendix 3   Health and safety

Where possible avoid going out alone to record mammals. If you do go out on your own, always ensure that someone knows where you are going and when you will be back.

The following notes are for information only; they are not a complete First Aid Manual! Always take care to avoid problems and, whenever an incident occurs, seek qualified assistance as soon as possible.

Bites and stings

Bites and insect stings are rarely fatal but occasionally cause an allergic reaction. Symptoms of such an allergic reaction include anxiety, widespread red blotching skin eruption, swelling of the face and neck, puffiness of the eyes, impaired breathing and rapid pulse and people exhibiting these symptoms should be taken to hospital immediately. If someone is bitten or stung the principal aim should be to reassure them and, if necessary, arrange for their removal to hospital.

In severe cases, the casualty should be kept as immobile as possible and the affected part kept below the level of the heart. Where possible a description of the animal that caused the problem should be given to those treating the patient.

Diseases which may affect mammal workers in the countryside

The following diseases can affect workers in the countryside. Please read this and take action as required. If you are handling small mammals and scats/spraint you should be especially aware of these. Avoid these activities if you have open cuts and grazes, or wear surgical gloves, and always wash your hands with detergents afterwards.

Tetanus

All outside workers should be protected with a tetanus injection, and anyone carrying out conservation work should check with their doctor that they have an up-to-date tetanus jab.

Lyme disease

Lyme disease is a bacterial infection transmitted through the bite of ticks. These ticks are generally found attached to vegetation or various mammalian hosts, particularly sheep and deer, although cattle and dogs have also been known to be infected. Ticks are particularly active between April and October and during this period you must be aware of the possibility of picking up a tick from vegetation, both for yourself and for pets.

Prevention

- ❏ Wear long trousers tucked into socks.
- ❏ Wearing light coloured clothing can help spot the ticks.
- ❏ Use insect repellent on your clothes and on pets.
- ❏ Brush off your clothes before entering buildings.
- ❏ Check for ticks when undressing as they can stay in clothing for hours before attaching to you.
- ❏ Remove any ticks attached to you by tugging at the mouth parts with tweezers. Alternatively smother them in detergent or olive oil. Do not try to remove them by pulling on the round body sac.
- ❏ Keep any removed ticks for identification purposes.

Diagnosis of Lyme disease and treatment

Infection may cause red rashes or patches from 3 - 45cms across and normally ring shaped. Flu-like symptoms are also common. More rarely, meningitis-like symptoms may occur, such as a stiff neck, difficulty in concentrating and general tiredness. If, after being bitten by a tick, you suffer from any of the above, see a doctor. Tell him/her that you have been bitten by a tick and may be infected with Lyme disease. Lyme disease can be treated with antibiotics at any stage, but the earlier it is diagnosed, the easier it is to treat.

 ! If in any doubt consult your doctor

Blue-Green algal blooms

Blue-green algae occur naturally in many inland waters. In still water in the summer they can multiply quickly and turn the water green, blue green or green, brown. During calm weather the algae may rise to the surface causing a scum to form. Toxins may also be released by the algae. In humans the illness resulting from swallowing or coming into contact with infected water may be severe, but there are no reports of long term effects. Pets are more likely to suffer a severe illness, and death may result.

Prevention

❏ Avoid both algal scum and the water around it.

❏ In areas where scum is present make sure that pets are not allowed to approach the water, and in particular, ensure that they do not drink the water.

Diagnosis and treatment

Illnesses resulting from algal blooms include skin rashes, eye irritation, vomiting, diarrhoea, fever and pains in muscles and joints. If you believe that you are infected because you show some of the above symptoms, contact your doctor immediately and tell him that you may have come into contact with algal scum.

Similarly, if you believe that your pets may be infected, contact a vet and inform him that they may have come into contact with an algal bloom.

Leptospirosis

Leptospirosis in the U.K. occurs mainly in two types. One form is Weil's disease which occurs in the urine of various wild animals, particularly rats, and from them can pass into water systems. The second, the Hardjo form of Leptospirosis, occurs in cattle and can be passed to humans. Incidences of Weil's disease are very rare, but the infection can cause kidney failure and thereby death. For conservation workers the most likely source of Weil's disease is from stagnant water which has been infected.

Prevention

❏ When in contact with stagnant water or soil that may have been contaminated, always wear boots and gloves.

❏ Cover all cuts and broken skin with waterproof plasters before and after work.

❏ Ensure that water does not get into the eyes, nose or mouth. Do not bite your nails.

❏ After working with or near dirty water, wash your hands and forearms with soap and water, particularly before eating, drinking or smoking. Equipment should also be rinsed and dried as soon as possible.

❏ Avoid contact with rat urine.

Diagnosis and treatment

Symptoms begin within 3 to 19 days of being exposed to the bacteria. Flu-like symptoms are common, including a high temperature and muscle pains. Other symptoms include conjunctivitis and jaundice.

If you suffer from any of the above symptoms and have possibly been exposed to Weil's disease, contact your doctor immediately. Explain that you may have been exposed to the disease and ask for an 'ELISA' blood test to check for the presence of the disease. Early diagnosis and treatment are vital for recovery.

For more information contact the Health & Safety Executive (www.hse.gov.uk) who run an information line and produce a leaflet on Leptospirosis.

Appendix 4   Key to the identification of small mammal skulls

First, are you sure it is a small mammal skull? Note that amphibian skulls are very flat and have no teeth. Reptile skulls fall apart easily and are therefore found as fragments. Their teeth are minute, almost too small to see.

Bat, Insectivore or Rodent?

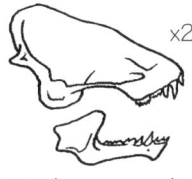

BAT (PIPISTRELLE)

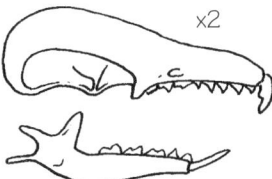

INSECTIVORE
SKULL LONG AND NARROW, NO
PROMINENT CHEEKBONES. TEETH
EVENLY SPACED ALONG JAW.

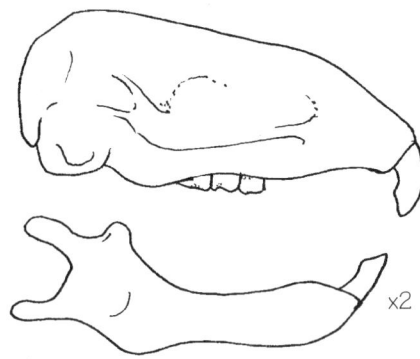

RODENT
SKULL BROAD WITH PROMINENT CHEEKBONES. BIG
GAP BETWEEN INSCISORS AND CHEEK TEETH

A. Insectivores

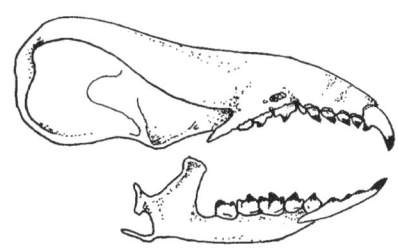
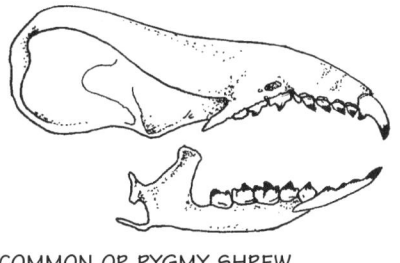

COMMON OR PYGMY SHREW

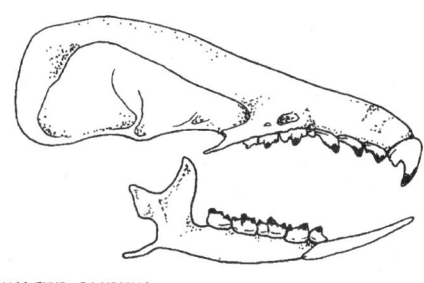

WATER SHREW

COMMON SHREW
Skull(21 mm)   _____
Lower jaw (14 mm)   _____
PYGMY SHREW (APPEARS VERY TINY)
Skull(12 mm)   _____
Lower jaw (10 mm)   _____

WATER SHREW
Skull(22 mm)   _____
Lower jaw (16 mm)   _____

> **Shrew teeth are tipped
> with red unless badly worn.**

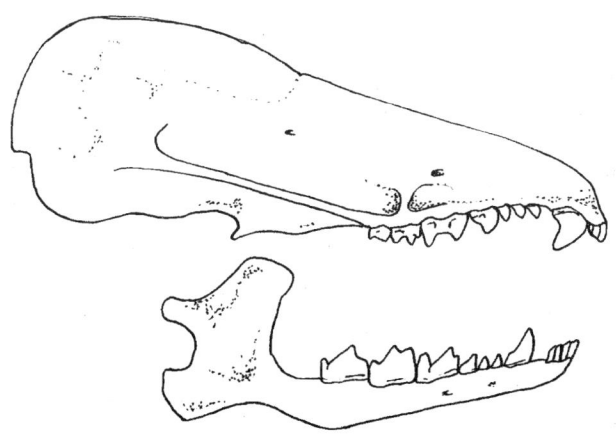

MOLE
VERY LARGE SKULL. TEETH ARE WHITE,
LOWER INCISORS SMALL, NOT LONG BLADES
AS IN SHREWS.

Skull   _____
(34 mm)
Lower jaw   _____
(23 mm)

B. Rodents

Voles can be distinguished from mice and rats by the shape of the tooth sockets. Pull out the front molar from the upper jaw and examine the hole left behind.

Voles

FIELD AND BANK VOLES ARE A SIMILAR SIZE BUT THEIR TEETH DIFFER. LOOK AT THE UPPER TOOTH ROW:

incisors at front

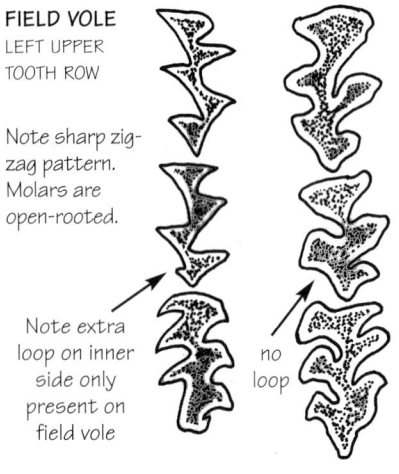

FIELD VOLE
LEFT UPPER TOOTH ROW

Note sharp zig-zag pattern. Molars are open-rooted.

Note extra loop on inner side only present on field vole

BANK VOLE
LEFT UPPER TOOTH ROW

Note rounded pattern, especially old animals. Molars have roots.

no loop

VOLE TEETH LEAVE ONE LONG RAGGED SOCKET WHEN REMOVED.

LOWER JAW

Empty socket with tooth removed

Teeth of upper and lower jaws have a bold zig-zag pattern.

VOLE TEETH

BANK VOLE (MOLARS HAVE ROOTS)

 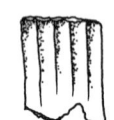 

Note young animals won't have developed roots yet

roots developing

old bank vole - molars have prominent roots

FIELD VOLE (MOLARS OPEN-ROOTED)

(Water Vole is simlar)

Molars grow throughout the life of the animal

AVERAGE SKULL SIZES:

FIELD/BANK VOLE
Skull(23 mm)
Lower jaw (13 mm)
Tooth row (9 mm)

WATER VOLE (MUCH LARGER)
skull (43 mm)
lower jaw (32 mm)
tooth row (12 mm)

Mice

TEETH LEAVE SEVERAL HOLES FOR INDIVIDUAL TOOTH ROOTS.

Teeth have knobbly surfaces and separate roots.

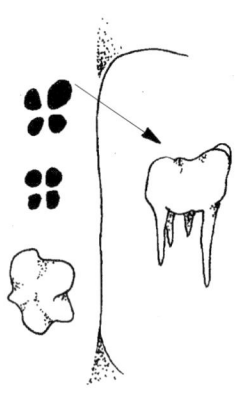

HAZEL DORMOUSE
4 CHEEK TEETH WITH SEVERAL PARALLEL TRANSVERSE RIDGES

ROOTS OF THE FIRST UPPER MOLAR (M1) DIFFER (M1= THE ONE NEAREST THE INCISOR)

UPPER JAW	LOWER JAW
(M1)	M1 to M3

Incisors at front

HOUSE MOUSE
3 ROOTS
Skull (24 mm)
Lower jaw (15 mm)

 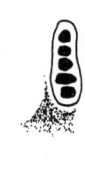

WOOD MOUSE
4 ROOTS
Skull (24 mm)
Lower jaw (15 mm)

note characteristic small socket

HARVEST MOUSE
5 ROOTS
Skull (16 mm)
Lower jaw (10 mm)

 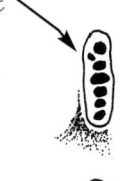

RAT
Skull (36 mm)
Lower jaw (25 mm)

 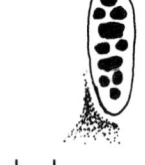

RULER

5 10 15 20 25 30 35 40 45 50
mm

Note: drawings are not to scale.

Illustrations by A. Beer & J. Savery with guidance from A. Love.

Appendix 5 How to identify water shrew faeces

Size approximately 7mm by 2mm; when intact, noticeably larger than common shrew and pygmy shrew faeces

Colour dark brown/blackish when fresh (becoming paler when dry) with shiny fragments of exoskeleton

Texture granular, easily crumbled between fingers (in comparison, rodent faeces are tough and fibrous)

While the water shrew eats many terrestrial prey items it is the only small mammal to prey on aquatic invertebrates, which form the bulk of its diet. An easy way of detecting the presence of this shrew in aquatic habitats is to look for the presence of aquatic remains in its faeces. To do this, place in a petri dish, add a few drops of water and gently crush to aid separation of fragments. Then carefully scan the dish using a binocular microscope (x50 magnification) for any aquatic prey items. Most common are remains of freshwater shrimps, water slaters and caddis larvae (see diagrams below). A mass of whitish, opaque exoskeleton is also a clear indication of freshwater shrimp.

Diagrams of diagnostic prey remains *(from Barber and Churchfield in prep).*

FRESHWATER SHRIMP - Gammarus

Two or three evenly sized and shaped segments followed by a relatively long and thin tarsus with a curved claw. These features are very indicative of a Gammarus leg.

WATER SLATER - Asellus

Highly variable and uneven segment shapes and sizes dependent on the function of the leg and its position along the body. Distinguishable from terrestrial isopods, such as woodlice, by the absence of large spines on the tarsi.

CADDIS FLY - Various Species

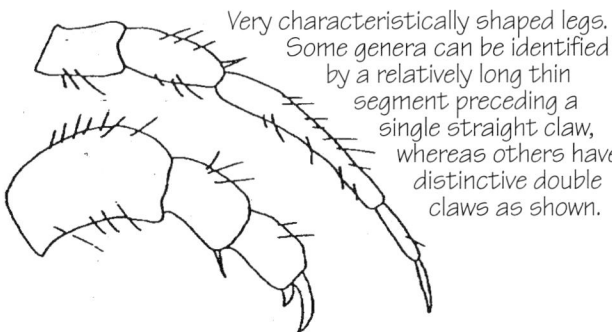

Very characteristically shaped legs. Some genera can be identified by a relatively long thin segment preceding a single straight claw, whereas others have distinctive double claws as shown.

Illustrations by J. Barber

Listed below are references, providing detailed diagrams and additional notes, to help familiarisation with water shrew prey items.

Croft, P.S. (1986). A Key to the Major Groups of British Freshwater Invertebrates. AIDGAP 181. Reprinted from Field Studies 6 No. 3 Field Studies Council, Preston Montford. ISBN 1 85153 181 6.
Macan, T.T. (1974). A Guide to Freshwater Invertebrate Animals. Longman. ISBN 0 582 32275 X.
Tilling, S.M. (1987). A Key to the Major Groups of British Terrestrial Invertebrates. AIDGAP 187. Reprinted from Field Studies 6 No. 4. Field Studies Council, Preston Montford. ISBN 1 85153 187 5.

Form I **General mammal recording form**

N.B. Each record MUST contain information under headings in bold

Observer's code			Name		Address		Tel		
Date	**Grid Ref**		Locality (or site name)	**Species**	Habitat*	Abun-dance*	Search method*	Breeding status*	Activity*
	2 letters	6 figures							
01.01.97	SP	35 205	1/4m SW of Grange Farm	Harvest mouse	J23 (Hedge)	1	XN (Nest)	-	-

Notes:

1. This record card has been completed and returned on the understanding that the information provided will be entered into a computer database and will be used for nature conservation, research and education.
2. Contact your County Mammal Recorder to be allocated an Observer Code.
3. * = Refer to The Mammal Society code card page 68 for codes and abbreviations.
4. More than one code for Abundance, Search method and Breeding status can be used on each record.

Please return form to your county mammal recorder, or, if unknown to: The Mammal Society.

How to Find and Identify Mammals
The Mammal Society

Form 2 **Site recording form**

N.B. Each record MUST contain information under headings in bold

Observer's code	Name	Address		Tel

Grid Ref		Locality	Habitat*	
2 letters	6 figures	(or site name)		
SP	435 205	1/4m SW of Grange Farm	J23 (Hedge)	

Date	Species	Abundance	Search method	Breeding status	Activity	Notes
01.10.98	Harvest mouse	1	TL (Longworth)	–	–	

Notes:

1. This record card has been completed and returned on the understanding that the information provided will be entered into a computer database and will be used for nature conservation, research and education.
2. Contact your County Mammal Recorder to be allocated an Observer Code.
3. * = Refer to The Mammal Society code card for codes and abbreviations.
4. More than one code for Abundance, Search method and Breeding status can be used on each record.

Please return form to your county mammal recorder, or, if unknown to: The Mammal Society.

THE **Mammal** SOCIETY

Form 3 **Owl pellet recording form**

N.B. Each record MUST contain information under headings in bold

Species of owl	**Date** of collection	Sample no.
Grid Ref (2 letters 6 figures)	Locality	**Name** of collector
Habitat	Site of roost	

Prey species (Give number of each)

Pellet no.	Checked OK	Length mm	Breadth mm	M.ag	Cleth	Apod	Micromys	Mus	Rattus	S.aran	S.min	Neomys	Bird	Other 1	Other 2	Min no. prey
1																
2																
3																
4																
5																
6																
7																
8																
9																
10																
11																
12																
Min no																

Notes:

1. This record card has been completed and returned on the understanding that the information provided will be entered into a computer database and will be used for nature conservation, research and education.
2. Contact your lCounty Mammal Recorder to be allocated an Observer Code.
3. * = Refer to The Mammal Society code card for codes and abbreviations.
4. More than one code for Abundance, Search method and Breeding status can be used on each record.

Please return form to your county mammal recorder, or, if unknown to: The Mammal Society.

N.B. Each record MUST contain information under headings in bold

Grid Ref of line or grid		Day of session	**Date**
Total traps	Spacing		
Habitat*			
Weather since last trap round: clear/cloudy; day/night; wet/dry; hot/warm/mild/cold/below zero			
Further notes			
Observer's **name**			

Trap no.	**Species**	New or retrap	New no	Sex m/f	Reproductive condition	Weight g	Notes
35	Woodmouse	R		F	GR	23	

√delete as appropriate

M.ag	M.ar	Cleth	A.flav	A.sylv	Micromys	Mus	Rattus	S.aran	S.min	Neomys	Other	TOTAL

Summary: number captured by species for all traps (excluding recaptures)

Notes:
1. This record card has been completed and returned on the understanding that the information provided will be entered into a computer database and will be used for nature conservation, research and education.
2. Contact your County Mammal Recorder to be allocated an Observer Code.
3. * = Refer to The Mammal Society code card for codes and abbreviations.
4. More than one code for Abundance, Search method and Breeding status can be used on each record.
Please return form to your county mammal recorder, or, if unknown to: The Mammal Society.

Recording Form Code Card

code	description	code	description	code	description
A1	Woodland	E22	Mire: Flush/spring, basic	I11	Rock: Natural exposure, inland cliff
A11	Woodland: broadleaved	E23	Mire: Flush/spring, bryophyte dominated	I12	Rock: scree
A12	Woodland: coniferous	E3	Fen	I13	Rock: limestone pavement
A13	Woodland: mixed	E31	Fen: valley mire	I14	Rock: other natural exposure
A2	Scrub A21 Scrub: dense/continuous	E32	Fen: basin mire	I15	Rock: Natural cave
A22	Scrub: scattered	E33	Fen: flood plain	I16	Rock: Mountain top
A3	Parkland	E4	Mire: bare peat	I17	Rock: Riverine
A31	Parkland/scattered trees: broad-leaved	F1	Swamp, fen	I19	Rock: other basic rock exposure
A32	Parkland/scattered trees: coniferous	F11	Swamp: single sp. dominant swamp	I1A	Natural rock exposures/caves excl. limestone
A33	Parkland/scattered trees: mixed	F12	Swamp: tall fen vegetation	I2	Rock: artificial exposure
A4	Recently felled woodland	F2	Marginal/inundation	I21	Rock & waste: artificial exposure, quarry
A41	Recently felled woodland: broad-leaved	F21	Marginal/inundation: marginal	I22	Rock & waste: artificial exposure, spoil heap
A42	Recently felled woodland: coniferous	F22	Marginal/inundation: inundation	I23	Rock & waste: artificial exposure, mine
A43	Recently felled woodland: mixed	G1	Standing water	I24	Rock & waste: artificial exposure, refuse-tip
B	Grassland	G11	Open water: standing, eutrophic	J	Cultivated
B1	Grassland:acidic	G12	Open water: standing, mesotrophic	J1	Cultivated/disturbed land
B11	Grassland: acid, unimproved	G13	Open water: standing, oligotrophic	J11	Cultivated/disturbed land: arable
B12	Grassland: acid, semi-improved	G14	Open water: standing, dystrophic	J12	Cultivated/disturbed land: amenity grassland
B2	Grassland:neutral	G15	Open water: standing, marl	J13	Cultivated/disturbed land: ephemeral
B21	Grassland: neutral, unimproved	G16	Open water: standing, brackish	J14	Cultivated/disturbed land: introduced shrub
B22	Grassland: neutral, semi-improved	G2	Open water: running water	J2	Boundaries
B3	Grassland:basic	G21	Open water: eutrophic running water	J21	Boundaries, intact hedge
B31	Grassland: calcareous, unimproved	G22	Open water: mesotrophic running water	J22	Boundaries, defunct hedge
B32	Grassland: calcareous, semi-improved	G23	Open water: running, oligotrophic	J23	Boundaries, hedge with trees
B4	Grassland: improved	G24	Open water: running, dystrophic	J24	Boundaries, fence
B41	Grasslnad: improved/reseeded, upland	G25	Open water: running, marl	J25	Boundaries, wall
B42	Grassland: improved/reseeded, lowland	G26	Open water: running, brackish	J26	Boundaries, dry ditch
B5	Grassland: marsh/marshy grassland	H1	Coastal	J27	Boundary removed
B51	Grassland: marshy, upland	H11	Coastland: intertidal, mud/sand	J28	Boundaries, earth bank
B52	Grassland: marshy, lowland	H12	Coastland: intertidal, shingle/cobbles	J3	Built up area
B6	Grassland: poor semi-improved	H13	Coastland: intertidal, boulders/rocks	J31	Built up area, agricultural/forestry
C1	Tall herb	H14	Coastland: intertidal, Zostera beds	J32	Built up area, industrial
C11	Tall herb and fern: Bracken, continuous	H15	Coastland: intertidal, green algal beds	J33	Built up area, domestic
C12	Tall herb and fern: Bracken, scattered	H16	Coastland: intertidal, brown algal beds	J34	Built up area, caravan site
C2	Tall herb and fern: upland species-rich ledges	H1A	Coastland: Intertidal mud and sand	J35	Built up area, sea wall
C3	Tall herb and fern: other tall herb or fern	H2	Coastland saltmarsh	J36	Built up area, buildings
C31	Tall herb and fern: other, tall ruderal	H21	Coastal: saltmarsh, Spartina dominated	J4	Bare ground
C32	Tall herb and fern: other, non-ruderal	H22	Coastal: saltmarsh, non-Spartina dominated	J5	Other habitat
D1	Heathland	H23	Coastland: saltmarsh/dune interface	K	Marine
D11	Heathland: dry dwarf shrub heath, acid	H24	Coastland: saltmarsh, scattered plants	RA	Woodland, plantation & scrub
D12	Heathland: dry dwarf shrub heath, basic	H25	Inland saltmarsh	RB	Grassland and marsh
D2	Heathland: wet dwarf shrub heath	H26	Coastland: Saltmarsh, dense/continuous	RC	Tall herb & fern
D21	Heathland: wet dwarf shrub heath, upland	H3	Coastland: shingle above high tide mark	RD	Heathland
D22	Heathland: wet dwarf shrub heath, lowland	H4	Coastland: boulder/rocks above high tide mark	RE	Mire
D3	Heathland: Lichen/bryophyte heath	H5	Coastland: strandline vegetation	RG	Open water
D31	Heathland: Lichen/bryophyte heath, upland	H6	Coastland: sand dune	RH	Coastlands
D32	Heathland: Lichen/bryophyte heath, lowland	H61	Coastland: fore dunes	RJ	Other (Arable, urban & artificial)
D4	Heathland: montane heath/dwarf herb	H62	Coastland: yellow dunes	YRV	Roadside Verge
D5	Heathland: dry heath/acid grassland mosaic	H63	Coastland: grey dunes	ZBL	Building
D51	Heathland: dry heath/acid grassland mosaic	H64	Coastland: dune slack	ZBR	Bridg
D52	Heathland: dry heath/acid grassland mosaic	H65	Coastland: dune grassland	ZBX	Bat Box
D6	Heathland: wet heath/acid grassland mosaic	H66	Coastland: dune heath	ZCH	Church
D61	Heathland: wet heath/acid grassland mosaic	H67	Coastland: dune scrub	ZCV	Cave
D62	Heathland: wet heath/acid grassland mosaic	H68	Coastland: open dune	ZFO	Fortification
E1	Bog, flush	H7	Coastal lagoons	ZHR	House roof
E11	Mire: blanket bog	H8	Coastal: maritime cliff and slope	ZHT	House Hanging Tiles
E12	Mire: upland raised bog	H81	Coastland: hard cliff	ZIH	Ice House
E13	Mire: lowland raised bog	H82	Coastland: soft cliff	ZKL	Kiln
E14	Mire: valley bog	H83	Coastland: maritime cliffs, crevice vegetation	ZMI	Mine
E15	Mire: basin mire	H84	Coastland: coastal grassland	ZTL	Road/Rail Tunnel
E17	Mire: wet modified	H85	Coastland: coastal heathland	ZTR	Tree
E18	Mire: dry modified	H86	Coastland: bird cliff vegetation	ZTW	Waterway Tunnel
E2	Mire: Flush/spring	I	Rocks		
E21	Mire: Flush/spring, acid/neutral	I1	Natural rock exposure		

SEARCH		SEARCH		BREEDING		ACTIVITY	
code	description	code	description	code	description	code	description
AB	aural bat detector	TN	trapped in mist net	AD	adult	CY	colony
BO	bottled	TR	trapped (other)	BA	Breeding possible	D	dead
BOC	canned	XB	skelatal remains	BC	Breeding confirmed	HI	hibernating
C	casualty (not road)	XD	feeding remains	BP	Breeding probable	Q	alive
CT	caught (not trapped)	XDU	dung/droppings	F	female	QTO	in torpor
DOR	dead on road	XH	hair/fur	GR	pregnant female		
O	olfactory record	XMH	mole hill	IM	immature		
QNB	nest box	XN	nest	IMF	immature female		
QSE	seen	XO	burrow/nesthole	IMM	immature male		
QTN	tin	XP	bird pellet	J	juvenile		
R	reported to recorder	XS	skin	M	male		
SH	shot	XT	tracks/trail	NC	nursery colony		
TB	trapped in breakback trap			QFL	lactating female		
TE	trapped in emergence trap			QIP	imperforate female		
TL	trapped in Longworth trap						